My First Encounter with ASTRONOMY

Leonardo Martinez

Kendall Hunt
publishing company

Kendall Hunt
publishing company

www.kendallhunt.com
Send all inquiries to:
4050 Westmark Drive
Dubuque, IA 52004-1840

Copyright © 2009 by Kendall Hunt Publishing Company

ISBN 978-0-7575-6259-4

All rights reserved. No part of this publication may be reproduced,
stored in a retrieval system, or transmitted, in any form or by any means,
electronic, mechanical, photocopying, recording, or otherwise,
without the prior written permission of the copyright owner.

Printed in the United States of America
10 9 8 7 6 5 4 3 2 1

Contents

Introduction vii

UNIT ONE—

Chapter 1: Exploring the Heavens 3

    Observing the Sky at Night 3

    The Celestial Sphere 5

    The Motion of the Earth 8

    Angles in the Sky and the Metric System 11

    Motion of the Moon 13

    Sidereal and Synodic Months 16

    Eclipses 16

    Parallax 19

    Appendix 20

    Answers 21

Chapter 2: A Brief History of Astronomy 23

    Early Greek or Classical Astronomy 23

    Modern Astronomy and the Heliocentric
Model of the Solar System 27

    Tides 35

    Conclusion 38

    Answers 38

Chapter 3: Interaction of Light with Matter 39

    Nature of Light 39

    The Electromagnetic Radiation and
Electromagnetic Spectrum 41

    Earth's Atmosphere and the Electromagnetic Radiation 43

    Differences between Temperature and Heat 44

    Temperature Scales 44

    Radiation Laws 45

    Atoms and Spectra 48

    Light Spectra 51

    Emission Nebula 55

    The Doppler Effect 55

    Answers 58

Chapter 4: Telescopes 59
- Optical Telescopes 59
- Large Optical Telescopes 61
- Special Instruments in Telescopes 62
- Powers of a Telescope 63
- The Seeing Conditions of the Atmosphere and Adaptive Optics 67
- Optical Interferometry 68
- Observatories 69
- Radio Telescopes 69
- Near Infrared Telescopes 71
- Space Astronomy 72
- Hubble Space Telescope 73
- Answers 74

UNIT TWO—

Chapter 5: The Solar System 77
- Introduction 77
- The Main Characteristics of the Planets of the Solar System 78
- General Characteristics of the Terrestrial Planets 78
- General Characteristics of The Jovian or Giant Planets 80
- Escape Speed 82
- The Dwarf Planets 85
- Magnetic Fields of Planets and Stars 85
- Space Debris 87
- Formation of the Solar System 94
- The Formation of Planets 97
- The Greenhouse Effect on Earth and Global Warming 101
- Clearing the Nebula 102
- The Solar Nebula Hypothesis and the Properties of the Solar System 103
- Other Solar Systems 104
- Answers 105

UNIT THREE—

Chapter 6: Important Properties of the Stars 109
- Measuring the Distance to the Stars 109
- Luminosity and Apparent Brightness 113

Relation between Apparent Brightness and
the Distance (d) to the Stars 113

Surface Temperature of the Stars..116

Chemical Composition of Stars..118

The Hertzsprung and Russell (H–R) Diagram.....................118

Luminosity Classes of the Stars 123

Binary Stars and Mass Determination of Stars....................125

Answers 130

Chapter 7: The Lives of the Stars: From Birth to Main Sequence Stars 133

The Interstellar Medium 133

Molecular Clouds and Star Formation 138

Characteristics of T-Tauri Stars 143

Stars Spend Most of Their Lives as
Main Sequence...144

Nuclear Energy Generation in
Main Sequence Stars..146

Energy Transport Inside the Stars 148

A Brief Look at Our Star: The Sun 149

Star Clusters 151

Answers 154

Chapter 8: Death of the Star 155

Low Mass Stars or Red Dwarfs 155

Medium or Intermediate Mass Stars 156

Postmain Sequence Evolution of a
High Mass Star..167

Cepheid Variables Stars as Distance Indicators 172

Answers 173

Chapter 9: Neutron Stars and Black Holes 175

Neutron Stars...176

Pulsars 178

Neutron Stars in a Binary System..181

Millisecond Pulsar 182

Escape Speed in Compact Objects and
Black Holes...183

Approaching a Black Hole 186

Gamma Ray Bursts 188

Black Holes in Binary System 188

Answers 189

UNIT FOUR—

Chapter 10: The Milky Way Galaxy and Other Galaxies 193

- The Dimensions of the Milky Way 194
- Motion of the Stars in the Galaxy 199
- Age of the Milky Way 199
- The Mass of the Galaxy and Dark Matter 200
- Origin of the Milky Way 202
- Other Galaxies 203
- Distance Determination to Galaxies 206
- The Expansion of the Universe 208
- Superclusters of Galaxies and Large-Scale Surveys 210
- Galaxy Interactions and Collisions 211
- Active Galactic Nuclei 213
- Quasars and Gravitational Lensing 216
- A Unified Model of Active Galactic Nuclei 216
- Galaxy Formation in the Early Universe 217
- Answers 219

UNIT FIVE—

Chapter 11: Brief Introduction to Cosmology 223

- The Cosmological Principle 223
- Why Is the Night Sky Dark? 223
- The Future of the Universe 225
- The Cosmic Microwave Background 227
- Fluctuations of the CMB and the Formation of Galaxies ... 229
- Origin of the Elements 230
- Answers 232

Appendix A 233

Appendix B 235

Appendix C 237

Appendix D 241

Appendix E 243

Appendix F 245

Appendix G 259

Index 261

A Few Words of Introduction

The goal of this course is to convince you that it is worthwhile to study the universe. We are not isolated in the universe, rather we are part of the cosmos. The elements that our bodies are made of originated inside stars billions of years ago.

Since the early dawn of civilization, mankind have struggled to render an intelligible view of the universe. The earlier models of the universe were geocentric, i.e., having the Earth as the center of the solar system. This point of view prevailed until the 17th century when Copernicus challenged it with important scientific arguments in favor of a heliocentric (Sun-centered) solar system. However, it took about 200 years and the work of several scientists for his ideas to be fully accepted.

The models to explain the universe are based on direct observations. If we do not have the appropriate instruments, oftentimes our observations will not be accurate, and our explanations of the physical world may be incorrect. However, as long as we are willing to keep observing and improve our technology, we can progressively improve our understanding of the physical phenomena. In astronomy, this is the path that has been taken. It is referred to as the "scientific method." The scientific method is the backbone of science.

Astronomy is a science. Astronomy is interested in finding an understanding of the universe. We have come a long way, but there are many mysteries ahead of us, such as the existence and origin of dark matter and dark energy.

We will begin our journey by looking at the night sky. This will lead us to study the laws that govern the solar system. Then, we will ask ourselves how did the Sun and the planets originate? After answering this question, we leave the solar system to study the stars. We will see that the stars are born, have a youthful stage, a period of old age, and then they die. The next step will lead us to study our galaxy and other galaxies. Finally, we will have a glimpse at the origin of the universe itself.

I hope that you enjoy this journey, so when you contemplate a starry night you may admire with awe the creation in front of you.

This work would have not been possible without the support of the Chairman of the Physics department of Florida Atlantic University.

I dedicate this notes to the memory of my parents, to my wife, sons, daughter, my brothers, sisters and to my students at FAU.

UNIT 1

Unit one is covered in the first four chapters. The material studied in this unit lays down the foundation for the rest of the course. I encourage you to read these chapters diligently, because we will draw upon them frequently later in the course.

The first chapter provides a quick glance at the night sky as seen from Earth. Chapter one gives a short overview of the history of astronomy. Chapter two will explain the laws of physics needed to understand certain astronomical phenomena, such as star brightness. And, finally, chapter three includes a brief description of the telescopes that astronomers use.

I recommend you also take seriously the questions and problems that you find at the end of each chapter.

CHAPTER 1

Exploring the Heavens

Throughout history, people of all civilizations have been fascinated by the night sky. In this chapter, we are going to look at the dome of the sky and explain the motion of the stars, the Earth, and the Moon. We begin by looking at the night sky.

OBSERVING THE SKY AT NIGHT

It is best to observe the night sky away from city lights and on a clear, Moonless night. When we observe the night sky several times during the same night, we see that the stars and planets rise in the east and set in the west. This gives the impression that they "revolve around a motionless Earth." This observation lead earlier civilizations to conclude that the Earth is the center of the universe, around which everything revolves. This model to explain the universe is known as the **geocentric** model of the universe. See Figure 1-1. The geocentric model prevailed until the 17th century, when it was replaced by the Sun-centered universe or heliocentric universe.

The Constellations

Different stars are grouped into constellations. These are arbitrary groups of stars in a particular place in the sky.

Most of the names of the constellations were assigned by earlier civilizations.

The stars in a constellation are at different distances. They also have different sizes, masses, and brightness. The only relation between the stars in a constellation is that they are visible in the same region of the sky, otherwise they are unrelated to each other. See Figure 1-2.

Officially, the stars have been grouped into 88 constellations.

Within a constellation, the brightness of the stars is indicated with Greek letters: α is the brightest, β the second brightest, γ the third, δ the fourth, and so on. This brightness of the stars in the constellation Scorpio is shown in Figure 1-3.

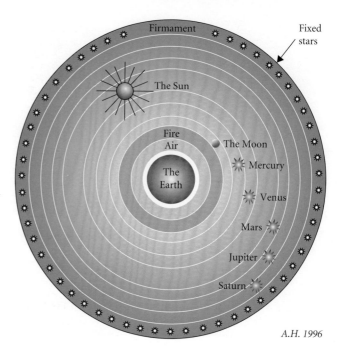

FIGURE 1-1. *The Greek astronomers' geocentric model of the universe.*

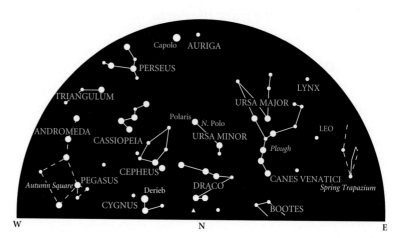

FIGURE 1-2. *Some of the constellations visible in the northern hemisphere.*

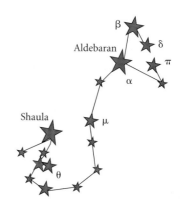

FIGURE 1-3. *The brightest star in the constellation Scorpio is Aldebaran, indicated by the letter α.*

Chapter 1—*Exploring the Heavens*

FIGURE 1-4. *The Big Dipper is one of the best known asterisms.*

Credit: NASA Johnson Space Center – Earth Sciences and Image Analysis (NASA-JSC-ES&IA)

Asterism

An **asterism** is a group of stars that is not officially a constellation. Sometimes the stars in an asterism are part of one or two constellations. For example, the Big Dipper (with seven stars) is part of the constellation Ursa Major or the Big Bear. See Figure 1-4.

THE CELESTIAL SPHERE

On a clear night, if you observe the stars for at least 30 consecutive minutes, you can see that the stars move. The stars that are low on the eastern horizon move higher and higher and new ones appear, while the stars near the western horizon move "down," and finally disappear below the horizon. Further observation will show you that the stars, constellations, and planets "rise in the east and set in the west." Based on these and other observations, Plato (350 BC) proposed that the celestial bodies, planets, Moon, and stars should move with constant speed in circular obits around the immobile Earth. This could be easily explained if the celestial objects were fixed on the internal surface of a rotating, transparent sphere called the **celestial sphere**. See Figure 1-5.

The celestial sphere is an imaginary, transparent sphere centered on the Earth that contains the stars and other celestial objects.

When the celestial sphere is viewed from Earth, it seems to rotate from east to west. This apparent rotation is caused by the rotation of the Earth on its axis from west to east. Since the celestial sphere carries all the celestial bodies, the Sun, the planets, and the stars seem to rise in the east and set in the west.

The model that has the Earth as the center of the universe is known as the geocentric model of the universe.

Important Elements of the Celestial Sphere

The celestial sphere seems to rotate around a fixed axis called the **celestial axis**. The celestial axis has two fixed pivots of rotation: the north and the south celestial poles (NCP & SCP), which are located above the North and South poles of the Earth

Q1. What is a constellation?
a. The set of all stars that belong to a galaxy
b. A pattern formed by stars that are in the almost same direction of the sky when viewed from Earth
c. A cluster of stars that were formed together and are close to each other in space
d. A group of stars that are all at the same distance from Earth

Q2. What is the name of the constellation shown below?

Q3. What is the name of the asterism shown below, and which star is the second brightest?

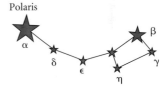

Q4. Identify the constellation shown below. Which star is the brightest?

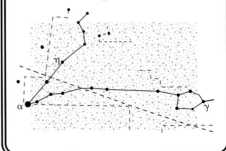

Chapter 1—*Exploring the Heavens*

Q5. Why do the stars move across the sky throughout the night?
a. Because the stars rotate around their axes
b. Because the Earth revolves around the stars
c. Because the stars revolve around the Earth
d. Because the Earth rotates around its axis
e. b & d

Q6. How many constellations are there (officially) in the sky?
a. 100 billion b. 880
c. 2,200 d. 120
e. 88

Q7. What is the name of the star at the NCP?
a. The Polar Bear
b. Alpha Centauri
c. Polaris
d. Betelgeuse

Q8. If you are at the North Pole, what do you see at the zenith?
a. The constellation Orion
b. The Sun
c. The celestial equator
d. Polaris

Q9. If you are at the Earth's equator, what do you see at the zenith?
a. The constellation Orion
b. The Sun
c. The celestial equator
d. Polaris

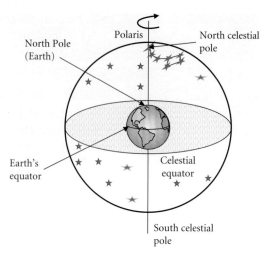

FIGURE 1-5. *Earth is the immobile center of the celestial sphere. This is the geocentric model of the universe.*

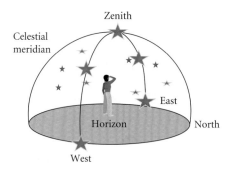

FIGURE 1-6. *The horizon and the zenith.*

as shown in Figure 1-5. The celestial axis is an extension of the Earth's axis. By luck, there is a moderately bright star, **Polaris**, positioned almost at the NCP. (There is not any star close to the SCP.)

The **celestial equator** is the imaginary line that divides the sky into two hemispheres. The celestial equator is an extension of the Earth's equator, so they are in the same plane.

The horizontal plane that separates the ground from the sky is the **horizon**. The stars rise above the eastern horizon and set below the western horizon. See Figure 1-6. If your right hand points east and your left hand points west, you will face north. The point directly overhead is called the **zenith**.

From any location on Earth, we only see the part of the sky that is above our horizon; the part of the sky below the horizon is invisible.

What you see at the zenith depends on your position on the Earth. If you are at the North Pole, Polaris will be at the zenith, but if you are at the equator, Polaris is in the horizon and the celestial equator is overhead. See Figure 1-7.

The number of stars that we see depends on our latitude on the Earth. If you are at the equator, latitude zero, you are able to see stars in both hemispheres. You see the maximum number of stars possible. As you move north or south from the equator, the number of stars that you can see diminishes. When you are at the South or North Pole, you would see the minimum number of stars possible.

When you are close to the North or South Pole, you will see that the stars in constellations around the SCP and NCP move in a cicle around the celestial poles. They never rise and never set. These are the **circumpolar constellations**.

The star Polaris, called the North Star, is very close to the NCP. Polaris does not move. If you were in the North Pole during winter, you would see the constellations shown in Figure 1-8 the entire night as they moved in a circle around Polaris.

A similar situation happens in the SCP.

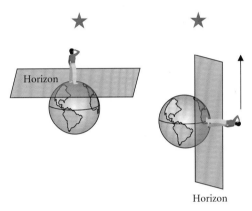

FIGURE 1-7. *The horizon in two different places on Earth.*

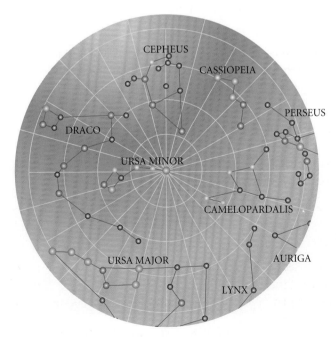

FIGURE 1-8. *Circumpolar constellations observed by a person very close to the North Pole.*

Chapter 1—*Exploring the Heavens*

The Motion of the Earth

The Earth has three different types of motion: rotation, orbital motion around the Sun, and precession.

Rotation

The daily or a diurnal motion of the Earth on its own axis, in 24 h, produces the day and night.

Orbital Motion

As the Earth rotates on its axis, it slowly moves along its orbit around the Sun. This means that our view on the stars constantly changes. During a year, or 325.25 days, our line of sight to the Sun moves through a group of constellations located along an apparent line called the **ecliptic**. If you observe the Sun just before sunrise for a few months, you will notice that every month the Sun rises with a different constellation. For example around February 21, the Sun rises with the constellation Aquarius. In March, it rises with the constellation Pisces, and so forth.

The ecliptic is called an **apparent line** because it is the result of the Earth's orbital motion around the Sun. The Sun does not move. The ecliptic is also the projection of the Earth's orbit on the celestial sphere. That is, the plane of the ecliptic contains the plane of the Earth's orbit as shown in Figure 1-9.

The **zodiac** is the zone in the sky within about 18 degrees on either side of the ecliptic. Over 1 year, the Sun seems to move through the constellations of the zodiac.

The motion of the Earth causes the stars to rise about 4 min earlier each day.

The constellations that are in the same direction as the Sun are invisible. For example, from June through September, the constellation Orion is behind to the Sun and cannot be seen, but between January and April, it is opposite to the Sun and can be observed.

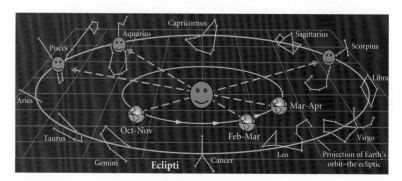

FIGURE 1-9. *As the Earth moves on its orbit around the Sun, the Sun seems to be moving across the sky though the constellations of the zodiac.*

Q10. What is the ecliptic?

a. The time at which an eclipse will occur

b. The point straight above us on the celestial sphere

c. The apparent line along which the Sun moves on the celestial sphere

d. The projection of the Earth's orbit on the celestial sphere

e. c & d

Q11. Refer to Figure 1-9, which constellation is not visible in October?

a. Virgo

b. Taurus

c. Aries

d. Sagittarius

Q12. Refer to Figure 1-9, which constellation is visible from October 21 to January 21?

a. Taurus

b. Scorpious

c. Sagittarius

d. Libra

The Earth's axis is inclined (tilted) 23.5 degrees with respect to its orbital plane, i.e., to the plane of the ecliptic. For this reason, the great circle that the ecliptic forms on the celestial sphere makes an angle of 23.5 degrees with the plane of the Earth's equator.

Figure 1-10 shows that half of the plane of the ecliptic is above the celestial equator and the other half is below it.

In the northern hemisphere, the point of the ecliptic where the Sun is at its most northern point above the celestial equator is the summer solstice. The summer solstice happens around June 21. The southernmost point below the celestial equator is the winter solstice, which happens around December 21. The two points at which the ecliptic intercepts the celestial equator are the equinoxes, as shown in Figure 1-10.

As the Earth revolves around the Sun, the inclination of its axis with the plane of rotation remains constant, and the amount of sunlight that the Earth receives changes as it moves in its orbit. This causes the seasons. See Figure 1-11.

In the northern hemisphere, the summer solstice marks the beginning of summer. This is the point at which the North Pole of the Earth is oriented toward the Sun. This happens around June 21. At this time in the southern hemisphere, the South Pole points away from the Sun, marking the beginning of winter in this hemisphere. Similarly, the winter solstice indicates the beginning of winter in the northern hemisphere.

The equinoxes at which the Earth crosses the celestial equator are the points of transition between summer and winter, as shown in Figure 1-12.

> **Q13.** The star Sirius in the constellation Canis Major rises tonight at 8:00 pm. At what time would it rise in 10 days?
>
> a. 7:20 pm
> b. 8:00 pm
> c. 8:20 pm
> d. 8:40 pm

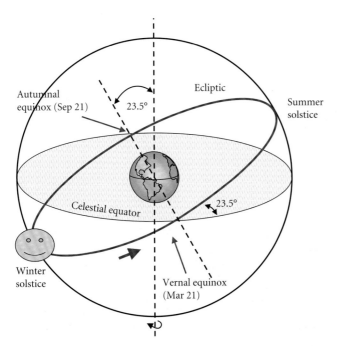

FIGURE 1-10. *The planes of the ecliptic and the celestial equator intercept each other at an angle of 23.5 degrees.*

Q14. When it is winter in Canada, it is _____ in Argentina.

a. also winter
b. summer
c. spring
d. fall

Q15. How do the zodiac constellations differ from other constellations?

a. Most of them are named after animals
b. They are located along the celestial equator
c. They are located in the northern hemisphere
d. The planets move within them
e. a & d

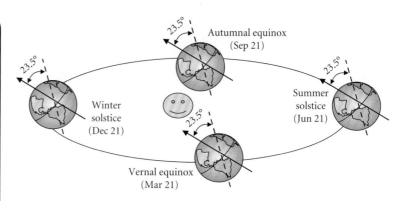

FIGURE 1-11. *The plane defined by the orbit of the Earth is the plane of the ecliptic.*

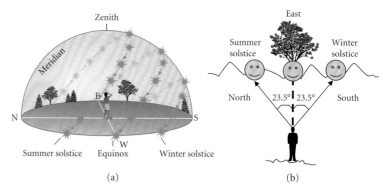

FIGURE 1-12. *Seasonal displacement of the path, rising, and setting of the Sun.*

Due to the tilt of the Earth and its motion around the Sun, the path that the Sun follows along the celestial sphere changes with each season. Also, the position of the sunrise and sunset on the celestial horizon is displaced as the seasons change. See Figure 1-12.

During the summer solstice, the Sun rises 23.5 degrees north of east and sets 23.5 degrees north of west.

During the equinoxes, the Sun rises due east and sets due west.

During the winter solstice, the Sun rises 23.5 degrees south of east and sets 23.5 degrees south of west. See Figure 1-12b.

The time it takes for a star to appear on the meridian for two consecutive nights is about 23 h and 54 min, this is called **sidereal day**. The time it takes for the Sun to appear at its highest in the sky for two consecutive days (the time from one noon to the next) is 24 h, or one solar day. The difference is caused by the fact that the Earth rotates on its axis as it orbits the Sun. See Figure 1-13.

The average time between two consecutive vernal equinoxes is 365.242 solar days. This is called a tropical year. To make things more complicated, the sidereal year, which is the time the Sun takes to return to the same position with respect to the stars, is a little bit longer. It is 365.256 solar days. The sidereal year is 20 min and 24 s longer than the tropical year.

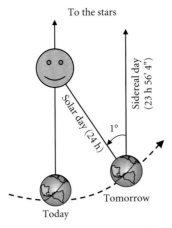

FIGURE 1-13. *Difference between solar and sidereal day.*

Earth: Precession

When we described the celestial sphere, it was pointed out that the celestial axis points toward the star Polaris. That has not always been the case, and it will not be the case in the future.

The combined gravitational attraction of the Sun and the Moon acting on the rotating Earth tends to twist the Earth upright. This causes the direction of the Earth's axis to slowly change and wobble with time. Four thousand years ago, the star Thuban, in the constellation Draco, was the North Star. In 12,000 years from now, the North Star will be the star Vega, in the constellation Lyrae. In about 26,000 years, 25,770 to be exact, the axis will point again to the star Polaris, completing a full cycle. See Figure 1-14. Hipparcus was the first person to observe the effects of precession.

It is believed that the orbit, the precession, and the inclination (tilt) of the Earth undergo small changes over very long periods of time. These changes affect the heat balance and modify the climate of the Earth. They might have been responsible for the ice ages, which seem to repeat in a cycle of about 240 million years. In 1920, the meteorologist Milutin Milankovitch suggested that the long-term changes in Earth's weather are linked to the changes of Earth's orbital shape, precession, and axial tilt.

ANGLES IN THE SKY AND THE METRIC SYSTEM

Astronomers measure distances across the sky using angles or angular separations. We measure angles in arc degrees, arc minutes, and arc seconds.

Arbitrarily, a circle is divided in 360 angles. Each of these angles is called a degree. A degree is divided in 60 smaller angles, or minutes, and a minute is further subdivided in 60 parts. Each of these parts is called a second.

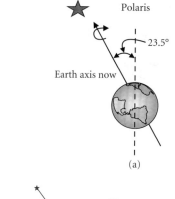

(a)

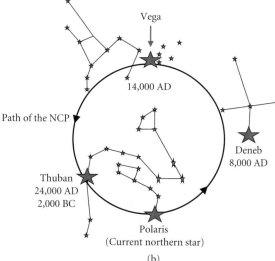

(b)

FIGURE 1-14. (a) *As the Earth precesses around the vertical, the axis forms a cone.* (b) *The circle shows the direction in the sky to which the axis of Earth will point as it precesses.*

Therefore,

1 arc minute = 1/60 of a degree
1 arc second = 1/60 arc of a minute
1 arc degree = 60 arc minutes = 3,600 arc seconds.

Figure 1-15 shows a practical way to measure some angular separations in the sky.

The angular separation, or angular diameter, **α**, of a celestial body is obtained by drawing two tangents to the opposite edges of the body as shown in Figure 1-16.

By coincidence, the Moon and the Sun have the same angular separation: **α = 0.5 arc degree**. This is because the proportion of their distance to their diameters gives the same value.

See the Appendix at the end of the chapter for more on this topic.

In astronomy, we use the metric system, which is based on the power of 10. For example, the primary unit of length is the meter (m) (about 3 feet).

The meter has multiples, such as the kilometer (km), and sub-multiples, such as the centimeter (cm) and the millimeter (mm).

Q16. We see different stars in summer than we see in winter because _____.

a. the Earth is at different points along its orbit

b. the brightness of the stars changes with the seasons

c. the stars are continuously moving in space

d. the Earth's axis is tilted with respect to its orbit

Q17. If you are at the Earth's equator, you will see the NCP _____.

a. in the zenith

b. in the horizon

c. 45 degrees toward the south

d. 45 degrees toward the north

Q18. 0.45 arc degrees is the same as ____ arc seconds.

a. 27

b. 270

c. 1,620

d. 60

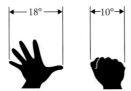

FIGURE 1-15. *A hand held at arm's length can be used to measure small angular distances.*

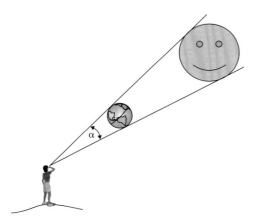

FIGURE 1-16. *The angular separation or angular diameter α of the Moon is equal to the angular separation of the Sun.*

1 km = 1,000 m = 10^3 m.
1 centimeter = 1 cm = 1/100 m = 0.01 m = 10^{-2} m.
1 millimeter = 1 mm = 1/1,000 m = 0.001 m = 10^{-3} m.
1 nanometer = 1 nm = 0.000000001 m = 10^{-9} m.

Other units of length that we will use are as follows:

1 AU = 150 million km = 150 × 10^6 km. (An AU is an astronomical unit.)
1 ly = 1 light year = distance that light travels in 1 year = ~ 10^{13} km.
1 parsec = 3.26 ly.

The primary unit of time is the second, and the primary unit of mass is the kilogram (kg).

MOTION OF THE MOON

The Moon moves around the Earth, and only one side of the Moon is visible from the Earth.

Why does the Moon always show the same features to us? Perhaps the Moon does not rotate on its axis as it circles the Earth. Observing the two diagrams of Figure 1-17, we conclude that the Moon does rotate on its axis as it progresses along its orbit around the Earth. Figure 1-17b shows that the Moon rotates once on its axis every time that it orbits the Earth once. The period of

> **Q19.** 0.45 arc degrees is the same as ____ arc minutes.
>
> a. 27
> b. 270
> c. 1,620
> d. 60
>
> **Q20.** 3,600 arc seconds is the same as ____ arc minutes.
>
> a. 27
> b. 270
> c. 1,620
> d. 60
>
> **Q21.** Due to the orbital motion of the Earth, the stars rise about 4 min earlier each day. If a star rises at 10:00 pm today, at what time did it rise 4 days ago?
>
> a. 9:56 pm
> b. 10:56 pm
> c. 9:00 pm
> d. 9:44 pm
> e. 10:16 pm

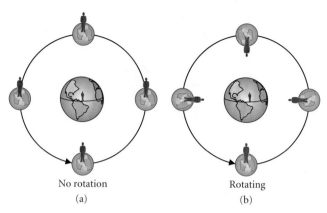

FIGURE 1-17. (a) *If the Moon did not rotate, we would see different features of the Moon throughout a month.* (b) *Because the Moon has synchronous rotation, we always see the same side of the Moon.*

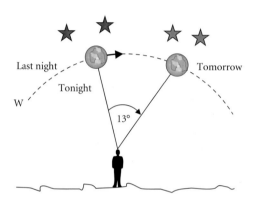

FIGURE 1-18. *In one night, the Moon moves about 13 degrees with respect to the stars.*

rotation is equal to the orbital period. This is called **synchronous rotation**.

The time for one rotation, or one revolution, around the Earth is 27.3 days. (This is called a **sidereal month** and it will be explained shortly.)

The Moon has a "near side," which is always visible from Earth, and a "far side," which is never visible. This is illustrated in Figure 1-17b.

The displacement of the Moon in the sky is also interesting. If you observe the Moon on several consecutive nights, you will discover that it moves about 13 degrees east (about 0.5 degrees per hour) with respect to the position of the previous night. See Figure 1-18. This explains why the Moon rises ~ 53 min later each day.

Lunar Phases

As the Moon moves on its orbit around the Earth, the amount of sunlight that it receives from the Sun changes gradually. Therefore, the Moon goes through different phases, as illustrated in Figure 1-19.

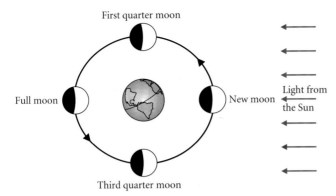

FIGURE 1-19. *Lunar phases.*

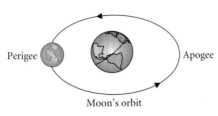

FIGURE 1-20. *The Moon moves around the Earth in an elliptical path.*

When the Moon is between the Earth and the Sun, the part of the Moon that faces Earth is not illuminated and it is difficult to see. This phase is called the **new Moon**. When the Moon is on the opposite side, it is fully illuminated and is called the full Moon. Between the new and full phases, the Moon goes through the first quarter Moon, and between full and new phases, it goes through the third quarter phase. The Moon goes through the four phases in 29.53 days. (This is called the **synodic month**.) Between each phase there is an interval of roughly 1 week.

Starting at new Moon, the lunar phases are first quarter, full, third quarter, and new Moon. The approximate rising and setting times of the Moon for each phase are given below.

PHASE	MOONRISE	MOONSET
New	Dawn	Sunset
First quarter	Noon	Midnight
Full	Sunset	Dawn
Third quarter	Midnight	Noon

A word of caution: The orbit of the Moon is tilted about 5 arc degrees with respect to the Earth's orbit. This means that the two orbits are not in the same plane. See Figure 1-22.

The orbit of the Moon is not a perfect circle; it is an ellipse with a small eccentricity, so the Moon does not move with constant velocity around the Earth.

The closest point of the Moon's orbit to Earth is called **perigee** and the farthest, **apogee**. See Figure 1-20.

> **Q22.** How many arc degrees per hour does the Moon move eastward with respect to the stars?
> a. 0.54
> b. 1
> c. 0.1
> d. 1.2
>
> **Q23.** At what time does first quarter Moon rise?
> a. At noon
> b. At dawn
> c. At midnight
> d. At sunset
>
> **Q24.** If the Moon rises tonight at 6:00 pm in 7 days it will rise at _____.
> a. 9:07 pm
> b. 10:24 am
> c. 11:11 pm
> d. 12:11 am

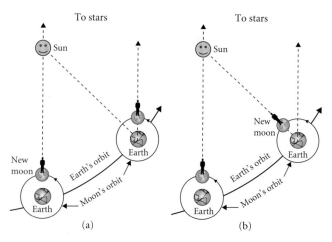

FIGURE 1-21. (a) *Sidereal month.* (b) *Synodic month.*

SIDEREAL AND SYNODIC MONTHS

As the Moon orbits the Earth, both the Earth and the Moon change position with respect to the Sun and the stars. The time the Moon takes to circle the sky once and return to the same position in the sky relative to the stars is the **sidereal month**. The sidereal month is equal to 27.32 days. See Figure 1-21a. During this time, the Earth moves on its orbit around the Sun, so it takes the Moon a little longer to return to the same position with respect to the Sun. This is the **synodic month**, and it is equal to 29.53 days. See Figure 1-21b. The length of our months (except February) are about 1 or 2 days longer than a synodic month.

ECLIPSES

Eclipses happen when the Moon moves into the shadow of the Earth or when the Earth enters the Moon's shadow, as shown in Figure 1-22.

The conditions for eclipses happen twice a year when the Moon intercepts the plane of the ecliptic. See Figure 1-22a and c.

In principle, eclipses should occur every new and full Moon. This is not the case, however, because the orbit of the Moon is tilted about 5 arc degrees with respect to the Earth's orbit, as shown in Figure 1-23.

Lunar Eclipse

A total lunar eclipse occurs at full Moon when the Moon enters the Earth's shadow, or **umbra**. Although the Moon is being eclipsed, it does not disappear completely. Some sunlight (mostly red light) makes it through the Earth's atmosphere to the Moon's surface. This happens due to refraction. The red light causes the

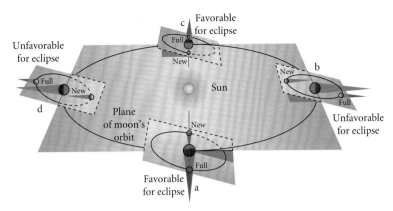

FIGURE 1-22. *Eclipses occur when the Moon, Earth, and Sun are aligned, as in a and c. This condition is met twice a year.*

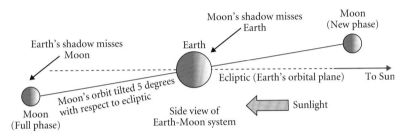

FIGURE 1-23. *The orbit of the Moon is inclined about 5 degrees with respect to the orbit of the Earth, for this reason we do not have eclipses every month.*

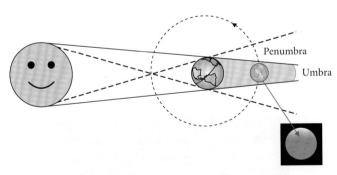

FIGURE 1-24. *A lunar eclipse happens during full Moon when the Earth is between the Sun and the Moon.*

copper-red color that the Moon adopts during total lunar eclipses. See Figure 1-24.

The maximum time of a total eclipse is about 1 h and 45 min.

If the Moon only enters the **penumbra**, the eclipse is partial.

Lunar eclipses are observed by the people who can see the Moon at the time of the eclipse, which is about half of the Earth.

Chapter 1—*Exploring the Heavens*

Solar Eclipse

A solar eclipse occurs when the Earth passes through the Moon's shadow. See Figure 1-25. If the Earth enters the umbra the eclipse is total, otherwise it is partial. The umbra is just 240 km wide, and only a few people on Earth can see the same solar eclipse.

The Moon's shadow moves at 1,700 km per hour, so a total solar eclipse lasts a maximum of 7.5 min, on average between 2 and 3 min.

During a total solar eclipse, the solar chromophere (lower red portion) and the solar corona become visible. See Figure 1-26.

Because the Moon's orbit around the Earth is not perfectly circular, if at the moment of the eclipse the Moon is at apogee, its farthest point away from the Earth, we see an annular solar eclipse. In this case, the Sun is not darkened much.

Roughly half of all total solar eclipses are annular. If the Earth only enters the penumbra, a partial solar eclipse comes into view. In Figure 1-27, the three types of eclipse are illustrated.

The Moon can occult the Sun because the angular diameter of both the Sun and the Moon is about the same, 0.5 degrees of arc.

The same eclipse is total for some people and partial for others. Why?

FIGURE 1-25. *Solar eclipses are observed when the Moon is between the Sun and the Earth. This occurs during or near a new Moon. See video at:* http://video.google.com/videoplay?docid=-4909789369249504644.

Q25. What is the phase of the Moon during a solar eclipse?

a. New or near new
b. Full or near full
c. Either full or new
d. Third quarter

Q26. If the orbits of the Earth and Moon were in the same plane, we would see eclipses every month.

a. True
b. False

Q27. For a lunar eclipse to happen, the phase of the Moon is _____.

a. new or near new
b. full or near full
c. either full or new
d. third quarter

FIGURE 1-26. *During a total eclipse the Sun's atmosphere becomes visible.*

Credit: NASA Goddard Space Flight Center (NASA-GSFC)

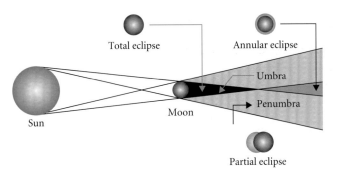

FIGURE 1-27. *Annular and partial solar eclipses.*

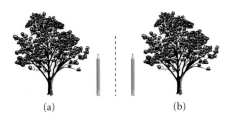

FIGURE 1-28. (a) *What you see with your right eye closed.* (b) *What you see with your left eye closed.*

Due to the mutual interaction between the Earth and the Moon the radius of the Moon's orbit grows about 3.8 cm every year. At the beginning of last century, the Moon was about 3.82 m (10 feet) closer to the Earth. In about 1.4 billion years, the solar eclipses will be only annular eclipses. (Why?)

Parallax

Having a pencil in your hand, stretch out your arm, and view a distant tree. Now close your right eye and then, without moving the pencil, change eyes. When you do this you see that the tree seems to move as shown in Figure 1-28. This apparent displacement of the object due to the change of position of the observer is called **parallax**.

If you repeat the previous experiment with objects located at different distances you might, I hope, conclude the following: Objects close to the observer have large parallax and objects far away have small parallax.

The Moon is relatively close to us, so it has a large parallax. The stars, which are far away, show very small parallax. A star's parallax can be only measured with telescopes.

When we take two observations of a constellation 6 months apart, we expect to see stellar parallax because we are observing the stars from two different points that are separated by 300 million km. However, naked eye parallax is not noticeable. The parallax is only observable with good telescopes. This is illustrated in Figure 1-29.

Q28. The apparent displacement of an object due to the displacement of the observer is _____.

Q29. Why isn't there an eclipse at every new and every full Moon?

Q30. Why is the Moon copper-red during a total lunar eclipse?

Q31. What phase would Earth be in if you were on the Moon when the Moon was full?

Q32. The region of partial eclipse is the _____ and the region of totality is the _____.

Q33. What astronomical phenomenon is related to the use of months in our calendar?

   a. The Earth's rotation around its axis

   b. The lunar phases

   c. The Earth's revolution around the Sun

   d. The planets' apparent motion on the celestial sphere

Q34. The sidereal month _____ the synodic month.

   a. is longer than

   b. is shorter than

   c. has the same length as

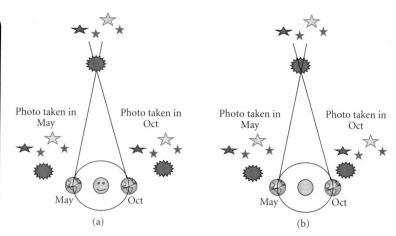

FIGURE 1-29. (a) *The stars do not show naked eye stellar parallax.* (b) *Telescope observations show that stars have parallax.*

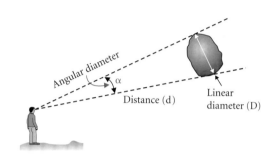

FIGURE 1-30. *The angular separation or angular diameter α of a celestial body.*

The Greek astronomers, and later the Europeans, did not see stellar parallax because they made their observations with the naked eye. For this reason, they concluded that the Earth was not moving and therefore it was the center of the universe.

Stellar parallax was measured for the first time in 1838, when telescopes were perfected.

The distance to the stars is measured in light years and in parsecs (pc). A **light year** (ly) is the distance that the light travels in 1 year. A **parsec** (pc) is equal to 3.26 ly. The closest star to us is the red dwarf Proxima Centauri, which is about 4.3 ly away.

Appendix

Angular separation or small-angle formula.

The angular diameter (**α**), the linear diameter (**D**) of a celestial body, and the distance (**d**) to Earth are related through the small-angle formula. See Figure 1-30.

$$\frac{\text{angular diameter }(\alpha)}{57.3°} = \frac{\text{Diameter }(D)}{\text{distance }(d)}$$

$$\frac{\alpha}{57.3°} = \frac{D}{d}$$

The angle α is given in arc degrees and the distance in km, AU, or even ly.

Example

The linear diameter of Jupiter is about 140,000 km. What is the angular separation of Jupiter when it is 780 million km away from Earth?

Solution Using the small-angle formula and solving for α, we have

$$\alpha = 57.3 \, (D/d) = 57.3 \, (140{,}000 \text{ km}/780{,}000{,}000 \text{ km}).$$
$$\alpha = 0.01 \text{ degrees} = 0.60 \text{ arc of minute} = 3.6 \text{ arc of a second}.$$

ANSWERS

1. b
2. Leo
3. The little dipper. The star identified with the Greek letter β is the brightest.
4. Pisces, the star α
5. d
6. e
7. c
8. d
9. c
10. e
11. d
12. a
13. a
14. b
15. e
16. a
17. b
18. c
19. a
20. d
21. d
22. a (13/24 =?)
23. a
24. d, The Moon rises 53 min later each day, so in 7 days it will rise 53 × 7 = 371 min = 6 h 11 min later, so it will rise at 12:11 am.
25. a
26. a
27. b
28. parallax
29. Because the orbits of the Earth and Moon are in slightly different planes.
30. Because the Earth's atmosphere refracts, or bends, the light from the Sun.
31. New
32. penumbra, umbra
33. b
34. b

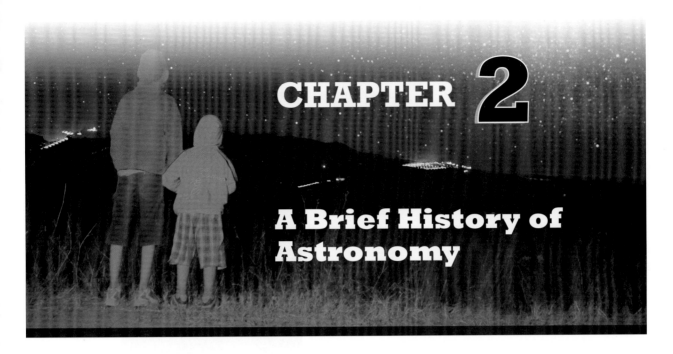

CHAPTER 2

A Brief History of Astronomy

In this chapter, we will present a short overview of the most important epochs in the history of astronomy.

Earlier civilizations were great observers of the skies. They built monuments to predict the time of the day, the seasons, and other astronomical events such as the equinoxes and solstices.

The ruins of the Mayan observatory in Central America, and the monument of Stonehenge in England, are witnesses of the attempts of previous civilizations to explore the sky. For example, if you stand at the center of the inner circle at Stonehenge during the summer solstice, you'll see the Sun rising directly over the "heel stone." This chapter is divided into three parts: the classical or pre-Copernican era, the Copernican revolution, and the modern and contemporary era.

EARLY GREEK OR CLASSICAL ASTRONOMY

The classical period of astronomy coincides with the development of the Hellenistic Greek period.

The following Greek thinkers made important contributions to astronomy: Pythagoras (580–500 BC), Eudoxus (400–347 BC), Plato (427–347 BC), Aristotle (384–322 BC), Aristarchus of Samos (310–230 BC), Eratosthenes (276–195 BC), Hipparchus (140 BC), and Claudius Ptolemy (140 AD).

Pythagoras believed that the Earth was round and that each planet was attached to its own crystalline spheres. He also concluded that the orbit of the Moon was inclined to the Earth's orbit.

Eudoxus was the first astronomer to construct a geometric model of a **geocentric universe**, with the stars located inside the huge rotating sphere around the Earth. He reasoned that the Moon is the closest celestial body to the Earth because it circles the Earth in about only 28 days. He also concluded that Saturn had to be the farthest celestial body from Earth because it takes roughly 29 years to circle the Earth. He was unaware of the existence of the planets beyond Saturn. Why?

Hipparchus (190–120 BC) was the first astronomer to observe the precession of the Earth and to classify the stars into six different classes according to their brightness or magnitudes. The concept of magnitude will be discussed in Chapter Six.

Some of the Greek intellectuals believed that the Earth was round, including Aristotle. Aristotle confirmed this common belief when he observed that the shadow the Earth casts on the Moon during a lunar eclipse is part of a circumference. See Figure 2-1. Eratosthenes later estimated Earth's circumference by measuring the distance between two cities and the angle the Sun rays made with the vertical, at noon during an equinox. (For a description of his experiment, refer to: http://www.astro.cornell.edu/academics/courses/astro201/eratosthenes.htm.)

FIGURE 2-1. *The shadow that the Earth casts on the Moon during a solar eclipse suggests that the Earth is round.*

Credit: NASA Solarsystem Collection

The Greeks developed the **geocentric model** to explain the universe based on naked-eye observations. This model depicts the celestial bodies rotating around the motionless Earth. The underlying principles governing the geocentric model of the universe were developed mainly by Aristotle (384–322 BC). His model was not based on experimental observations but on first principles. According to Aristotle, first principles are ideas that are obviously true and need not be proven.

The Aristotelian geocentric model of the universe is based on the following principles:

1. The Earth is the center of the universe and everything in the sky moves around the Earth from east to west. See Figure 2-2.

2. The planets, the Moon, the Sun, and the stars move around the Earth in perfect circles with constant speed, and each planet is always at the same distance from the Earth.

3. The heavens are perfect and immutable, and the Earth is imperfect and mutable.

4. All celestial objects are perfect spheres.

Aristotle's geocentric universe included 55 crystalline transparent nested spheres, turning at different but constant speeds and at

---

Q1. According to the Greek astronomer, Eudoxus, the Moon is closer to Earth than Jupiter because the Moon _____ .

a. looks bigger than Jupiter

b. has an atmosphere and Jupiter does not

c. circles the sky faster than Jupiter

d. is solid and Jupiter is not

Q2. The geocentric universe of nested spheres devised by the Greeks to explain the universe dominated astronomy for about 2,000 years.

a. True

b. False

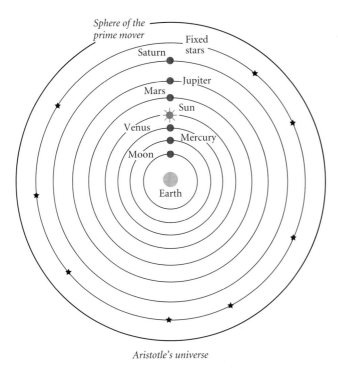

FIGURE 2-2. *Geocentric Greek universe.*

different angles. The stars were located inside an outer sphere; therefore, all the stars were at the same distance from the Earth.

Aristarchus (Ἀρίσταρχος) of Samos (310–230 BC) concluded that the Sun is the center of the universe. He is known as the **Copernicus of antiquity**. He also made careful records of the retrograde motions of the planets against the background of stars.

Aristarchus's contemporaries rejected the Sun-centered universe and adopted the geocentric model, because they did not see stellar parallax (see chapter one) and because the movement of the Earth was not felt.

In the geocentric model, the constant speed and the fixed distance at which the planets circle Earth are at odds with two important observations: 1) the change in brightness of the planets and 2) their retrograde motion.

The Greek astronomers were aware of the change in brightness of the planets, as they move "around the fixed Earth." If the planets are always at the same distance from Earth and are unchangeable, how is this possible? Furthermore, the astronomers observed the planets moving eastward with respect to the stars, but sometimes the planets appeared to stop and begin moving westward (**retrograde motion**) for a few weeks. The planets would then appear to stop again before resuming their normal motion (**prograde motion**).

The retrograde motion of planets is difficult to explain if the planets revolve around the Earth. Why?

> **Q3. The Greeks and other ancient civilizations were interested in observing the heavens because _____.**
>
> a. they knew that the Earth was revolving around the Sun
>
> b. the knowledge of the appearance of the heavens gave them information about the Earth's seasons
>
> c. they used the knowledge of the stars to navigate their ships
>
> d. they knew that the sky looked different from month to month because the Earth was moving in the plane of the ecliptic
>
> e. both b and c

Q4. The geocentric model of the universe proposed by Aristotle _____.

a. is an example of the scientific method
b. was based strictly on accurate observations
c. was based on first principles
d. both a and c

Q5. One of the first astronomers to suggest that the Sun, as opposed to Earth, existed as the center of the solar system was _____.

a. Aristotle
b. Ptolemy
c. Alexander the Great
d. Plato
e. Aristarchus of Samos

Q6. When the planet Mars moves westward in relation to the stars, it is _____.

a. in prograde motion
b. about to produce an eclipse
c. in retrograde motion
d. the beginning of summer

Q7. The geocentric model proposed by Aristotle gave a satisfactory explanation for the change in the brightness of Mars and the other planets.

a. True   b. False

Claudius Ptolemy of Alexandria (140 AD) created a mathematical model to explain the Aristotelian geocentric universe, incorporating the retrograde motion and the change in brightness of the planets.

In his model, the planets move in small circles, called **epicycles**, whose centers move in large circles centered on Earth, called the **deferent**, as illustrated in Figure 2-3.

Ptolemy's epicycles were necessary to explain the retrograde motion of the planets and their change in brightness. Every time the planet makes a loop, it is in retrograde motion and is closest to the Earth. At this point, the planet is at its brightest. When the planet moves out of the loop to resume its normal forward motion, its distance from Earth increases and its brightness diminishes.

In Ptolemy's model, the Earth was slightly off the center. The center of the epicycles moves only at a constant rate when observed from a point called the **equant**.

Ptolemy weakened the Aristotelian model in order to build a mathematical and quantitative model for the geocentric universe.

Ptolemy's model was not accurate, and in a few decades, the position of the planets did not correspond to the prediction of the model; therefore, the tables obtained with the model had to be revised every 50 years or so. Ptolemy was the last great classical astronomer, and his geocentric model of the universe prevailed for more than a thousand years, until the 16th century.

The Greeks were great astronomers eager to explain natural phenomena by way of reasoning in order to discover the basic laws of nature. Although experimentation was not their main concern, they were willing to test their ideas and to correct them.

The great respect and acceptance of the Aristotelian principles held back the advance of the experimental sciences for more than a thousands years. The advent of Copernicus brought about the departure of the Greek influence in science.

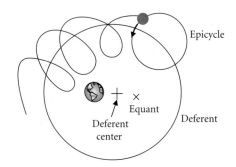

FIGURE 2-3. *The planet Mars moves on a small circle, or epicycle, and its center moves on a large circle, the deferent.*

Modern Astronomy and the Heliocentric Model of the Solar System

Modern astronomy begins with Nicolaus Copernicus (1473–1543) who, challenging the Aristotelian geocentric model, argued that the Sun and not the Earth was the center of the universe. This idea was further developed by Kepler, observationally confirmed by Galileo and fully explained by Isaac Newton in 1687 in his book *The Principia*.

> **Q8.** The Greek astronomers did not accept that Earth was revolving around the Sun because _____.
> a. none of them proposed that idea
> b. stellar parallax was not observed
> c. no motion could be felt
> d. their telescopes were not that good
> e. both b and c

The Sun-Centered Universe of Copernicus

Nicolaus Copernicus (1473–1543) rediscovered the Sun-centered system first proposed by Aristarchus of Samos. In his model, the planets, including the Earth, rotate on their axis and revolve around the Sun in perfect circular orbits and with constant speed. The planets move fastest when it is closest to the Sun and move slowest when it is farthest.

The Earth moves faster than the planets that are beyond it. Consequently, when the Earth periodically overtakes and passes these planets, they appear to move westward (retrograde motion) with respect to the stars, as illustrated in Figure 2-4. Since the distance from the Earth to the planet changes continually, the change in brightness of the planet follows.

The Copernican model was basically correct. However, it did not predict the position of the planets more accurately than the complicated but incorrect model of Ptolemy. The inaccuracy of the model arose in part because Copernicus erroneously assumed that the planets are moving in perfect circles with constant speeds as

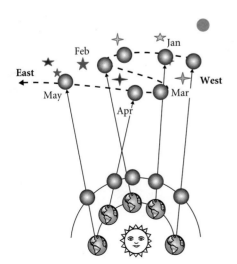

FIGURE 2-4. *When the Earth passes Mars faster, it seems to move with retrograde motion.*

> **Q9. In the Copernican model of the solar system, the planets moved on _____ orbits around the Sun.**
>
> a. parabolic
> b. circular
> c. elliptical
> d. hyperbolic
>
> **Q10. _____ show(s) retrograde motion.**
>
> a. The stars
> b. The Moon
> c. The planets
> d. Only Mars
>
> **Q11. The _____ in position of an object across the distant background when viewed from two different places is known as _____.**

required by the Aristotelian system. Copernicus was also forced to use epicycles, whose centers were moving on deferents, centered on the Sun rather than on the Earth.

The most important contribution of the Copernican model was to relegate the Earth to an undistinguished place in the solar system.

The Copernican heliocentric theory of the universe gave origin to the birth of modern astronomy and modern science.

The book, *De Revolutionibus*, published in the year of his death (1543), contained the fruit of his work. Only after his death, and based on new observations, his model was gradually accepted.

Copernicus' heliocentric model was rejected because

1. It was not more accurate than the well-known Ptolemy system and
2. Stellar parallax was not observed.

His model was simpler and more elegant than the complicated Ptolemaic model.

Tycho Brahe (1546–1601), a great Danish observational astronomer, collected a wealth of data about the motion of the Sun, Moon, and planets with respect to the position of the stars. Because he did not observe stellar parallax, he adopted, with interesting modifications, the geocentric view of the universe. In his model, the Earth was the stationary center around which the Sun and the Moon moved with the planets circling the Sun. See Figure 2-5. Brahe's measurements of the positions of planets on the sky were the most accurate and complete set of naked-eye measurements made up to his time. The accuracy of the position of the planets has only an error of 4 arc minute. This was great accuracy for naked-eye observations.

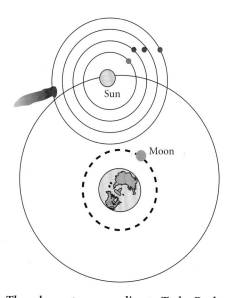

FIGURE 2-5. *The solar system according to Tycho Brahe.*

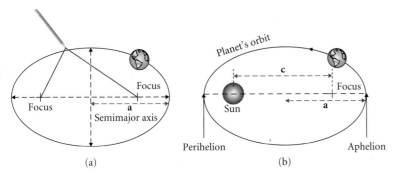

FIGURE 2-6. (a) *Drawing an ellipse with a pencil.* (b) *The Sun is at one of the foci of the ellipse.*

In 1792, Tycho observed a **supernova**. He also witnessed a comet in 1557. He proved that both the supernova and the comet were beyond the orbit of the Moon, in the realm of the heavens. These two discoveries proved that the skies were not unchangeable, as formerly explained by Aristotle. The Greeks thought that the comets happened in the atmosphere of Earth because they had accepted Aristotle's theory that the sky and the celestial objects were unchangeable.

Before we study Kepler's contributions to astronomy, we need to say a few words about ellipses, which are the shape of the orbits of the planets.

Figure 2-6 shows the main elements of an ellipse.

The shape of an ellipse is determined by the relative length of its horizontal or major axis and its vertical or minor axis. Half of the longer axis is called the **semimajor axis** and is usually represented by the letter **a**. When used to describe the motion of the planets, **a** is expressed in solar units (AU). (1 Au ~ 150 million km.)

An ellipse has two important points called the **foci**.

The eccentricity, **e**, is a number that tells us how much an ellipse departs from a circle. This number results by dividing the semimajor axis **a** by the distance between the two foci or **c**. A circle has an eccentricity of zero and the two foci concur at the center in one point. A very flat ellipse would have an eccentricity close to one, and the two foci are far away from each other.

Johannes Kepler (1571–1630) was a brilliant mathematician who worked with Tycho Brahe.

After Brahe's death, Kepler used the experimental data collected by Tycho and was able to show that the planets followed an elliptical orbit around the fixed Sun. With some modification, Kepler returned to the Sun-centered universe.

Planetary Laws of Johannes Kepler

1. Planets move on elliptical paths with the Sun at one of the foci.

2. The line from a planet to the Sun sweeps out equal areas in equal intervals of time (law of equal areas). If a planet takes the same time to move from **a** to **b** as from **c** to **d** (see Figure 2-7), then the two shaded areas are the same and the planet moves faster from **c** to **d** than from **a** to **b**.

3. The square of the orbital period **P** (in years) equals the cube of the semimajor axis of its orbit **a** (in AU). This is expressed in the following equation:

$$\mathbf{a}^3 = \mathbf{P}^2$$

The average distance of a planet to the Sun is equal to the semimajor axis **a**.

The orbits of the planets have small eccentricities and are virtually indistinguishable from circles. The most extreme examples are Pluto (with an eccentricity **e** = 0.248) and Mercury (with an eccentricity **e** = 0.206). Earth's eccentricity is only **e** = 0.0167.

Another interesting feature of the Keplerian model is that the closer a planet is to the Sun the faster it moves around it, as shown in Figure 2-8.

Application of Kepler's Laws

What is the average distance of Mars to the Sun if it has an orbital period of 1.88 years? (The period is measured by observing the planets as it circles the sky).

Solution   Using the previous equation, we find

$$\mathbf{a}^3 = \mathbf{P}^2, \mathbf{a} = \sqrt[3]{(1.88)^2} \text{ AU} = \sqrt[3]{(1.88)} \text{ AU} = 1.52 \text{ AU}$$

The average distance from Jupiter to the Sun is 5.2 AU. What is the period of time Jupiter takes to go around the Sun once?

$$\mathbf{a}^3 = \mathbf{P}^2, \mathbf{p} = \sqrt{(5.2)^3} \text{ years} = \sqrt{140.6} \text{ years} = 11.86 \text{ years}$$

Notice that the ratio of $\mathbf{a}^3/\mathbf{P}^2 = 1$ for all the planets of the solar system. Verify this rule for all the planets using Table 2-1.

Kepler expressed the distance from the Sun to the planets in terms of the **Earth–Sun distance** (astronomical unit) because he did not

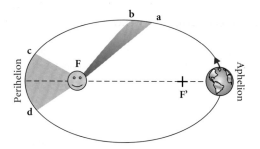

FIGURE 2-7. *The orbits of the planets are very similar to an ellipse. For illustration purposes, we have exaggerated the eccentricity.*

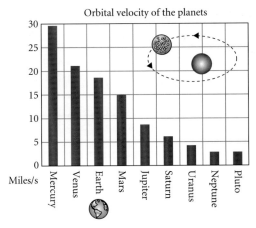

FIGURE 2-8. *The planets closer to the Sun move faster than the planets far from the Sun.*

Credit: © zoomschool.com

TABLE 2-1

PLANET	DISTANCE (AU)	PERIOD
Mercury	0.39	87.96 days
Venus	0.73	224.68 days
Earth	1	365.26 days
Mars	1.52	686.98 years
Jupiter	5.20	11.862 years
Saturn	9.54	29.456 years
Uranus	19.18	84.07 years
Neptune	30.06	164.81 years
Pluto (Dwarf planet)	39.53	247.7 years

know the value of the astronomical unit. The value of the astronomical unit was determined much later.

The heliocentric system was not easily accepted. Some of the objections were:

1. Common sense tells us that the heavens move around the Earth. If the Earth were moving forward, falling objects like stones, it was argued, would be left behind and we would feel a constant wind.

2. If the planets move on elliptical orbits, the speed of the planets will move faster when closer to the Sun and slower when far away. This effect should be noticed, but it is not. Therefore, the planets move in circular paths as predicted by Aristotle.

3. Stellar parallax is not observed; therefore, the Earth is at rest. (This was the most serious objection to Copernican model).

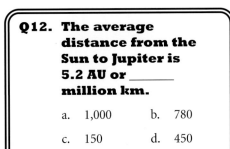

Q12. The average distance from the Sun to Jupiter is 5.2 AU or _____ million km.

a. 1,000   b. 780

c. 150   d. 450

> **Q13. A fictitious planet revolves around the Sun with a period of 27 years. What is its average distance to the Sun in AU?**
>
> a. 2 b. 3
> c. 6 d. 9
> e. 2.8
>
> **Q14. Tycho Brahe _____.**
>
> a. was known as a brilliant mathematician
> b. was convinced by astronomical observations that the planets moved around the Sun
> c. was an observational astronomer who collected and recorded lots of data about the position of the planets in the sky
> d. observed a supernova
> e. both c and d

Galileo Galilei (1564–1642)

Galileo was the most important experimental scientist of this time period. Legend prescribes that he used the leaning tower of Pisa to study the motion of falling bodies. He constructed his own telescope and used it to systematically observe the sky. He believed, along with Aristarchus of Samos, Copernicus, and Kepler, in the heliocentric system.

Galileo made the following astronomical observations. These observations conflicted with the Aristotelian model of the universe.

1. The Moon is not perfect but has mountains and craters. See Figure 2-9a. Aristotle had previously held that all celestial bodies were perfect.

2. Jupiter has moons orbiting it. See Figure 2-9b. Therefore, not everything revolves around the Earth as previously thought.

   (Actually Jupiter has many more moons. Galileo saw only the four largest moons. Today they are called the **Galilean Moons**.)

3. The Sun has sunspots; therefore, it is not perfect.

4. Venus shows a set of phases similar to the Moon's phases. Galileo was able to explain the occurrence of the phases only when he assumed that Venus revolves around the Sun (Figure 2-10).

These observations clearly supported the Copernican solar-centered solar system.

What keeps the planets moving around the Sun? Why do objects fall toward the surface of the Earth? Neither Kepler nor Galileo had the answer. The first person to correctly answer these questions was Sir Isaac Newton.

Newton's Universal Gravitation and Laws of Motion

Isaac Newton (1642–1727) was born on Christmas day in the same year Galileo died. When Newton was 25, he discovered the universal law of gravitational attraction and his three famous laws of motion.

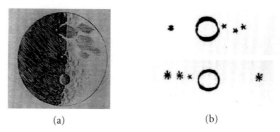

(a) (b)

FIGURE 2-9. (a) *One of Galileo's lunar drawings.* (b) *Galileo's drawings of the moons of Jupiter on successive nights.*

Chapter 2—*A Brief History of Astronomy*

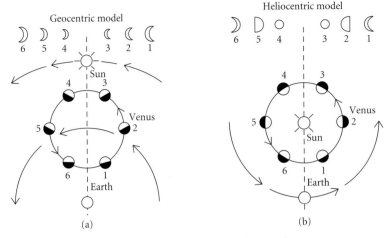

FIGURE 2-10. *Galileo's explanation of the phases of Venus.*

The universal law of gravitational attraction basically states that all bodies are attracted to each other no matter how far away they are from each other. Newton discovered that the force of attraction between two different objects of mass **M** and **m** is given by

$$F = G\frac{Mm}{d^2} \qquad (2\text{-}1)$$

In this equation, **G** is a constant, $G = 6.67 \times 10^{-11}$ Nm²/kg², **M** and **m** are the masses of the two bodies and **d** is the distance between their centers.

A simpler form of this equation is:

$$F \propto \frac{Mm}{d^2} \qquad (2\text{-}2)$$

∝ means proportional.

This equation states that the force of attraction between two bodies of mass **M** and **m** is directly proportional to the product of the two masses and is inversely proportional to the square of the distance between their centers. And this force is responsible for keeping the planets in orbit around the Sun.

This inverse square law states that the force between two masses varies as the inverse of the square of the distance between centers of the two objects.

For example, if the distance between two masses is doubled, the force between the masses decreases by one-fourth of the original value of force. This is because the square of the new distance is four times larger than the original distance. On the other hand, if we decrease the distance by one-half, the force will be four times larger. Why?

The "inverse square law" is very important in astronomy. Later on we are going to encounter other inverse laws.

Q15. An asteroid has a perihelion of 2.5 AU and an aphelion of 3.5 AU. The semimajor axis of this asteroid is _____ AU and its orbital period is _____ years.

  a. 6   14.6
  b. 4   8
  c. 3   5.2
  d. 2.8   4

Q16. If the distance between the Sun and the Moon were doubled, the mutual gravitational force would be _____.

  a. two times larger
  b. four times larger
  c. one-half of the original value
  d. one-fourth of the original value

Newton's laws of motion:

1. Law of inertia and mass
2. Force (**F**) = mass (**m**) × acceleration (**a**) or **F = ma**
3. Action force = Reaction force

The first law represents a major break-through from the Greeks, who believed that the natural state of things was to be at rest. Newton understood that, in order for a mass to change the direction of motion, it has to be acted upon by an external force **F**. Only by acting on an object of mass **m** with a force **F** can the object's direction or speed be changed.

The tendency of an object to keep its state of motion, at rest or moving with constant velocity, is known as **inertia**. The inertia is related to the mass, **m**, or amount of matter of a body.

For example, the force of gravity that the Earth exerts on the Moon continually changes the direction of the Moon's orbital velocity keeping the Moon in orbit. If the force disappears, the Moon would move on a straight line with constant velocity. See Figure 2-12.

Similarly the gravitational force that the Sun makes on the planets keeps them in orbit around the Sun. These are examples of bodies moving under the action of central forces.

Newton's second law states that the net force applied to an object is equal to the mass of mass, **m**, of the object times the resulting acceleration, **a**, according to the following formula,

Net force (F) = mass (m) × acceleration (a).

The gravitational force that the Sun applies to the Earth accelerates the Earth. This acceleration is what causes the direction of the

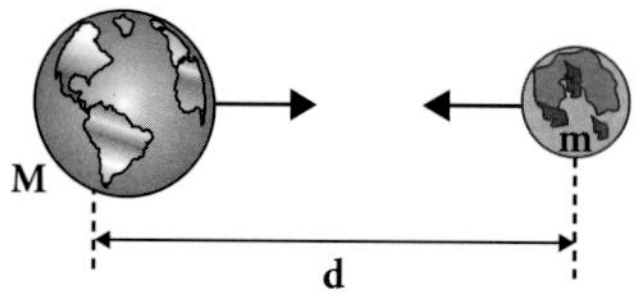

FIGURE 2-11. The Earth attracts the Moon and the Moon also attracts the Earth with an equal but opposite force.

> **Q17.** The gravitational force of attraction between two objects separated by a distance of 0.04 m is 10×10^{-6} N. If the distance between the two objects is reduced to 0.02 m, the new force will be _____ N.
>
> a. 40×10^{-2}
> b. 40×10^{-6}
> c. 2.5×10^{-6}
> d. 40

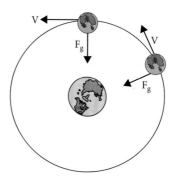

FIGURE 2-12. The force of gravity that the Earth exerts on the Moon keeps it in orbit around the Earth.

Earth's velocity to continually change its direction, so it can remain in orbit around the Sun.

When you kick a football, the force **F** of your foot initially accelerates the football and it gains some initial velocity under which it reaches certain distance, say 20 yards. Now, when you kick a lighter soccer ball, the same force of your foot acting on a much lighter mass gives a larger acceleration so the soccer ball reaches a larger distance, say 100 yards. However, if you kick the Great Wall of China nothing happens.

In general, the same force applied to objects produces different accelerations.

The third law states that to every action there is a reaction. For example, the Moon pulls on the Earth and the Earth, in response, pulls on the Moon with the same force.

The force of the Moon on the Earth also produces acceleration. However, the mass of the Earth is much larger than the mass of the Moon, so the motion of the Earth is not affected much by the force that the Moon exerts. A similar situation applies between the Sun and any of the planets.

Kepler's laws only apply to the motion of the planets in the solar system. Newton's laws, however, are valid anywhere in the universe and are applied to any object.

Newton was able to derive the three Kepler's laws using his laws of motion.

Newton's form of Kepler's third law gives a direct method for obtaining the masses of the planets, the stars, and even of the galaxies. For example, Newton determined that the mass of two stars rotating around a common center of mass, with a period **P**, which are separated by a distance **a** equals

$$(M + m) P^2 = a^3$$

This form of Kepler's third law is called **Kepler's modified third law**. In this equation, the masses are expressed in solar masses, the period in years, and the distance in AU. As we proceed, we will be using this form of Kepler's third law.

> **Q18.** The gravitational force of attraction between two objects separated by a distance of 0.02 m is 10×10^{-6} N. If the distance between the two objects is increased to 0.04 m, the new force will be _____ N.
>
> a. 40×10^{-2}
> b. 40×10^{-6}
> c. 2.5×10^{-6}
> d. 40
>
> **Q19.** You are pushing a wall with a force of 40 pounds; therefore, the wall pushes back on you with a force of 40 pounds.
>
> a. True b. False

TIDES

The tides are a consequence of the universal law of gravitation. The gravitational force, or tidal force, the Moon exerts on the Earth, even though weak, is strong enough to produce the tides. This force is a little stronger on the side of Earth facing the Moon than on the center, or on the far side, as shown in Figure 2-13. The consequence of this inequality is that the force pulling the oceans on the near side of the Earth is stronger than the force acting on the water on the far side. The water responds to this differential force by rising into a bulge of water on the side of Earth facing the Moon. There is also a smaller bulge on the other side of Earth due to the smaller force on this side. Therefore, the Earth is being pulled from under the water by the gravitational force. See Figure 2-13.

As the Earth rotates under the bulges, a given point on the shore experiences two tides (two low and two high) daily. The tides follow the Moon; therefore, high and low tides occur 53 min later each day. Remember that the Moon rises about 53 min later each day.

The gravitational pull of the Sun also contributes, to a lesser extent, to the high of the tides. During the first and third quarters, the Moon is at 90 degrees with respect to the Sun, and the gravitational force from the Sun is not in the same direction with the gravitational force of the Moon on the Earth. In this case, the gravitational force from the Sun will not influence much on the high of the tides (see Figure 2-14a). But during full or new Moon, the Sun, the Earth, and the Moon are aligned, and the gravitational force from the Sun has a larger effect on the high of the tides. See Figure 2-14b. In these cases, the low and high tides are bigger than during first and third quarter Moon.

The tides during first and third quarter are called **neap tides** and during full or new Moon are called **spring tides**. See Figure 2-14.

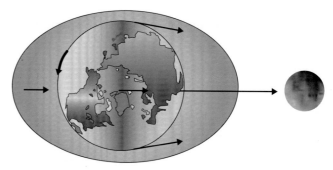

FIGURE 2-13. *The gravitational force, or tidal force, that the Moon exerts on the Earth causes bulges on two opposite sides of the Earth.*

FIGURE 2-14. (a) *Neap tides occur when the Moon is at 90 degrees with the Sun.*

FIGURE 2-14. (b) *Spring tides occur at new and full Moon when the three bodies are aligned.*

FIGURE 2-15. *Due to friction, the ocean bulge is not aligned with the Earth–Moon system. The mass of the bulge pushes the Moon outwardly.*

On average the depth of the tides is about 2 m (6 feet), but in some smaller and longer bays the tides reach a depth of up to 10 m (30 feet).

The Earth's crust is also affected a little bit by the pull of the Moon and can rise by 1 cm twice a day.

Some Consequences of the Tides

Friction between the ocean's water and the sea bed slows down Earth's rotation by approximately 1.5 ms per day per century (0.882 s per century). A day was only 18 h long 900 million years ago!

The Earth's rotation drags the ocean bulge ahead of the Earth–Moon center line. These tidal bulges have large amounts of mass, and they cause the Moon to move forward in its orbit, thus increasing the Moon's distance by about 3.8 cm/year. Nearly 1.2 billion years ago, the Moon was only 18,400 km from the Earth and a day was only 5 h long! At that time, the angular diameter of the Moon was also approximately 22 times larger than it is now. At present, the Moon is about 384,000 km away.

The Earth's days are getting longer, the Moon is receding, and the tides are getting weaker.

Also, Earth's tidal force flexes the rocks of the Moon, which subsequently produces friction. This has slowed down the Moon's rotation in such a way that now the Moon rotates once on its axis for each revolution around the Sun. This is called **synchronous rotation**.

If Moon is receding from the Earth 3.8 cm per year in 2,000 years, it will move only (3.8 × 2,000 =) 7.600 cm or 76 m. This is not a significant distance.

Q20. If the Earth had no Moon, then the tides would _____.
a. still occur but be barely measured
b. occur more often and with more intensity
c. would not occur at all
d. occur with the same frequency and with the same intensity

Q21. In a given place, the tides _____.
a. always happen at the same time
b. happen about 53 min later
c. happen about 53 min earlier
d. Always have the same intensity

Q22. If the Moon was one-half closer to Earth than it is now, the tides would be _____ intense as they are now.
a. as
b. twice as
c. four times as
d. eight times as

Q23. A major flaw in the formulation of Copernicus to explain the solar system was that he had _____.
a. the Earth moving around the Moon
b. the Earth at the center of the solar system
c. the planets moving with constant speed
d. the planets moving on almost perfect circles
e. both c and d

Chapter 2—A Brief History of Astronomy

Q24. An asteroid with an orbit lying entirely inside the Earth's orbit _____.

a. has an orbital period larger than 365 days
b. has an orbital period less than 365 days
c. has an orbital semimajor axis of less than 1 AU
d. has a highly eccentric orbit
e. both b and c

Q25. The Moon is slowly receding in its orbit.

a. True   b. False

Q26. Aristarchus (310–230 BC) of Samos, an island in the Mediterranean, was the first person to suggest that _____ was the center of the heavens.

a. the Earth
b. a large galaxy
c. the Sun and not the Earth
d. a star several times larger than the Sun

Conclusion

We have observed how the geocentric model of the universe was dethroned by the simpler model of the Sun-centered universe. We also see that new observations made at the beginning of the 20th century showed that this model wasn't correct either. The history of astronomy reveals that the scientific method is the method used in astronomy.

The scientific method depends on observation to support theories, models, and predictions.

A theory is a set of ideas used to explain certain experimental observations. A good theory allows us to make predictions about the behavior of nature. If the predictions are not accurate, more observations are made and the theory is refined. This is an ongoing process. This is the foundation of the scientific method.

Answers

1. c
2. a
3. e
4. c
5. e
6. c
7. b
8. e
9. c
10. c
11. apparent change; parallax
12. b
13. d
14. e
15. c
16. d
17. b, $[0.04/0.02 = 2]^2$ The distance is reduced by one-half; therefore, the force increases by a factor of 4.
18. c, The new distance is twice as much; therefore, the force is reduced by one-fourth.
19. a
20. a, The Sun also contributes to the intensity of the tides.
21. b, The tides follow the Moon and the Moon rises 53 min later each day.
22. c
23. e
24. e
25. a
26. c

CHAPTER 3

Interaction of Light with Matter

The light we receive from space brings information about the stars, galaxies, and other celestial objects. Therefore, it is imperative that we learn how to extract the information the light carries.

In this chapter, we will analyze the main properties of light, how it interacts with matter, and how we can interpret the information it carries.

NATURE OF LIGHT

When we say "light," we usually think about visible light. In astronomy, it has a broader meaning because it includes other regions of the electromagnetic spectrum such as ultraviolet (UV) and infrared light.

Several experimental results involving light, such as reflection, can be explained assuming that light is a wave. However, other results, such as the photoelectric experiment, can be explained if we assume that light consists of a stream of mass-less particle called **photons**. In general, when light interacts with matter, it behaves as a particle or a photon. When light propagates through space, it behaves as a wave.

The two theories of light are known as the **wave/particle duality** of light.

In 1860, James Clark Maxwell discovered the wave nature of light, and in 1902, Albert Einstein discovered the particle nature of light.

Before we can understand the general properties of the electromagnetic spectrum, it is necessary to define the terms associated with waves: **wavelength (λ), frequency (f),** and **period (P)**.

Wavelength, See Figure 3-1, is the distance from one wave crest to the next, and has units of length, the frequency **f**, (usually expressed in cycles per second or hertz) gives the number of oscillation per unit time. It is also the number of wave crests, or troughs, passing trough a given point per unit time. The period **P**, with units of time, is the duration of one oscillation.

The frequency **f** and the period **P** of a wave are related by:

$$f = 1/P \tag{3-0}$$

> **Q1. A neutron star has a frequency of 200 cycles per second. What is its period?**
> a. 10 s
> b. 0.001 s
> **c. 0.005 s**
> d. 5 s
>
> *Handwritten:* f = 200 cy/sec; f = 1/P; P = 1/f; P = 1/200; P = 0.005 sec
>
> **Q2. The distances between successive wave crests define the _____ of a wave.**
> a. frequency
> b. period
> **c. wavelength**
> d. an oscillation

Example

If the period of a wave is 0.01 s, then the frequency is 100 cycles per second, or 100 Hz, because 1/0.01 s = 100 cycles λ.

Frequency and period are inversely related since frequency increases as the period **P** decreases and vice versa. In astronomy, we frequently encounter this type of inverse relationship.

In a time equal to one period **P**, a wave front advances a distance of one wavelength, λ, thus the speed of the wave is equal to

$$\text{speed}(v) = \frac{\text{distance}(\lambda)}{\text{time}(P)} \qquad (3\text{-}1)$$

or

$$\text{speed}(v) = \text{distance}(\lambda) \times \text{frequency}(f)$$
$$v = \lambda \times f \qquad (3\text{-}2)$$

The speed of a wave is equal to its **wavelength (λ)** multiplied by its **frequency (f)**. We are most interested in the propagation of waves with the speed of light in a vacuum. The speed of light is always represented with the letter **c**, and in a vacuum, the speed of light is equal to 300,000 km/s. Replacing, in the last equation, **v** with the speed of light **c** we have

$$c = \lambda f. \qquad (3\text{-}3)$$

Wavelength and the frequency of light are inversely related, i.e., if the frequency increases, the wavelength decreases and vice versa. This is the same as:

$$f = c/\lambda. \qquad (3\text{-}4)$$

This inverse relationship is shown in Figure 3-2.

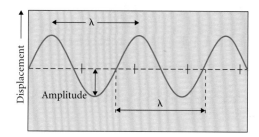

FIGURE 3-1. *Some characteristics of waves.*

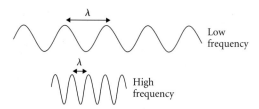

FIGURE 3-2. *The shorter the wavelength λ the higher the frequency of a wave.*

We often say that frequency and wavelength are related by

$f \propto 1/\lambda$.

The symbol $\propto$ means proportional.

Remember that light includes a broad spectrum of wavelengths called the **electromagnetic spectrum**.

THE ELECTROMAGNETIC RADIATION AND ELECTROMAGNETIC SPECTRUM

The energy emitted by the stars and other celestial object is called **electromagnetic radiation**. This electromagnetic radiation propagates in the form of electromagnetic waves.

Electromagnetic waves have a **varying electric** and **magnetic field** perpendicular to each other, as shown in Figure 3-3.

The same properties applicable to light also apply to electromagnetic radiation.

The wavelengths and frequencies of the most important regions of the electromagnetic spectrum are shown in Figure 3-4.

Notice that in Figure 3-4, the region of long wavelengths corresponds to low frequency, such as the radio radiation. The region of short frequency corresponds to high, such as X-ray and gamma radiation. This happens because frequency and wavelength are related through an inverse relation, as previously shown in equation 3-4. The electromagnetic radiation is important in astronomy because it carries the energy emitted by the stars, galaxies, and other celestial

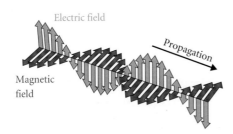

FIGURE 3-3. *Electromagnetic waves travel in a vacuum with a speed of 300,000 km/s (186,000 miles/s).* http://www.monos.leidenuniv.nl/smo/index.html?basics/light.htm.

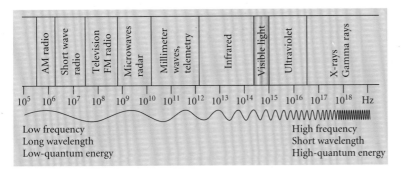

FIGURE 3-4. *Electromagnetic spectrum. Notice the inverse relationship between frequency and wavelength.*

$E \propto f \quad E \propto \frac{c}{\lambda}$

Q3. The _____ of an electromagnetic wave, the larger its energy.
   a. lower the frequency
   b. stronger the intensity
   c. longer the wavelength
   (d.) shorter the wavelength
   e. a and b

Q4. UV radiation carries _____ energy than infrared radiation.
   a. less
   (b.) more

Chapter 3—*Interaction of Light with Matter*

objects. This energy, **E**, is proportional to the frequency, **f**, and inversely proportional to the wavelength, **λ**, or

$$E \propto f \quad \text{or} \quad E \propto (c)/\lambda. \tag{3-5}$$

We see that a higher frequency (or a shorter wavelength) results in a larger energy **E**. Thus, the gamma rays are the most energetic, followed by X-rays. Radio waves are the least energetic because they have the longest wavelength.

Electromagnetic waves are fundamentally the same thing. They are waves of electromagnetic radiation traveling in a vacuum with the speed of light and carrying energy.

There is not a clear distinction between the different wavelength ranges. For example, the near infrared blends with the long-wavelength region of the visible spectrum.

The visible part of the electromagnetic spectrum is what we traditionally call light. The visible light is the collection of wavelengths to which our eyes are sensitive. The different wavelength of the visible spectrum are perceived as different colors. See Table 3-1. The collection of all visible wavelengths is perceived as white light.

TABLE 3-1. *Light of different frequency or wavelength is perceived with different color.*

FREQUENCY f(Hz)	WAVELENGTH λ(nm)	COLOR
7×10^{14}	400	Violet
		Blue
6×10^{14}		
	500	Green
5×10^{14}		Yellow
	600	Orange
4×10^{14}		
	700	Red

When white light passes through a glass prism, it is separated into its different wavelength components, as shown in Figure 3-5. Each color has its own range of wavelengths.

A glass prism separates the different colors because the index of refraction (bending) of glass is different for each wavelength or color.

Radiation of short wavelength is usually expressed in nanometer.

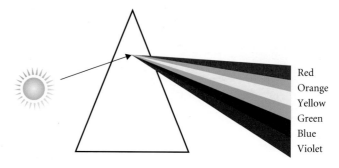

FIGURE 3-5. *The visible spectrum is obtained when white light passes through a prism.*

The infrared has two important regions. The near-infrared region refers to wavelengths between about 650 and 1,500 nm. The medium and far-infrared regions on the spectrum refer to longer wavelengths. The radiation emitted by stars and other objects in the infrared region is defined as **heat** or **thermal energy**. Wavelengths shorter than 400 nm reside in the near-UV region of the spectrum, followed by the mid- and far-UV region.

All the components of the electromagnetic spectrum behave as photons and display the **wave/particle duality** of light.

Photons have a frequency and wavelength, and photons carry energy.

As stated previously, the energy is proportional to the frequency:

$$E \propto f \qquad (3\text{-}6)$$

Thus, an X-ray photon carries more energy than a red photon.

Earth's Atmosphere and the Electromagnetic Radiation

Not all electromagnetic radiation coming from space reaches the surface of the Earth. The Earth's atmosphere acts as a filter. The atmosphere is transparent only to visible light and radio radiation. The atmosphere absorbs most of the UV and infrared light. Finally, it is completely opaque to X-ray and gamma rays. See Figure 3-6.

Thus, optical and radio telescopes are placed on the surface of the Earth, but X-ray and gamma ray telescopes have to be located above the Earth's atmosphere.

Q5. What is the best place to observe the sky using an infrared telescope?

a. From a high mountain top
b. From New York
c. From space
d. Cerro Tololo in Chile

Q6. As white light passes through a prism, the blue (shorter) wavelength bends less than the red (longer) wavelength. See Figure 3-5.

a. True
b. False

Q7. Observations in the X-ray portion of the electromagnetic spectrum are routinely done from the surface of the Earth.

a. True
b. False

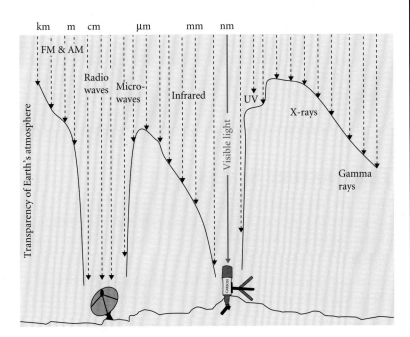

FIGURE 3-6. *The Earth's atmosphere discriminates the electromagnetic radiation that enters it.*

Chapter 3—*Interaction of Light with Matter*

> **Q8.** The average kinetic energy of the atoms of given object represents the average _____ of that object.
> a. mass
> b. heat energy
> (c.) temperature
> d. density

The stars are great sources of energy. Therefore, we must study the radiation laws. Before we do so, we must establish the difference between temperature and heat energy or heat. These terms are often mistaken as the same thing.

Differences between Temperature and Heat

Heat energy, or heat, and temperature are usually mistaken as the same thing, but there are many differences between them.

All bodies of mass are made of atoms and molecules.

The atoms of any object, regardless of its temperature, are constantly moving and twisting (jiggling). The average kinetic energy of the atoms or particles of a gas, liquid, plasma, or solid is the temperature. The total agitation, which is the sum of the kinetic energy of all the atoms or particles, is **thermal energy**. (The kinetic energy is the value of the mass of one atom, or particle, multiplied by the square of the average speed of each atom or particle divided by 2.)

At any temperature, the atoms and molecules of the gases, liquids, and solids are moving and they have heat energy.

At high temperatures, the atoms and molecules move quickly. Conversely, at low temperatures, the atoms and molecules move slowly.

Temperature is expressed in degrees. Heat is expressed in units of energy called **joules**.

The following example illustrates the difference between heat and temperature.

Consider a large iceberg and a hot cup of coffee. The trillions of atoms in the iceberg, on average, have small vibrations, thus low temperature. However, due to the great number of atoms, the heat energy associated with them is very large. On the other hand, the atoms in the cup of coffee are jiggling faster than the atoms in the iceberg and, therefore, have a higher temperature. The total heat associated with the coffee is less than the total heat of the iceberg because the iceberg has much more atoms than the hot cup of tea.

Temperature Scales

There are three different temperature scales in use today: Fahrenheit, Celsius, and Kelvin. Kelvin, or absolute temperature, is used in astronomy. Room temperature is about 300 K, and water freezes at 273 K.

The lowest temperature in the universe is about −459 °F, or 0 K. The Kelvin scale has no negative temperatures.

The relationship between the three scales of temperature is depicted in Figure 3-7.

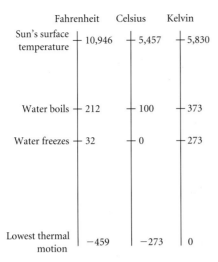

FIGURE 3-7. *Comparison between the three different temperature scales.*

The conversion formulae are

C to F: F = (C × 9/5) + 32

F to C: C = (F − 32) × 5/9   (3-7)

Example

Express 30 °C in F.

F = 30 × 9/5 + 32 = 54 + 32 = 86.

30 °C = 86 °F.

Express 92 °F in C.

The Kelvin and centigrade scales are related by

K = 273 + C.   (3-8)

RADIATION LAWS

Any object radiates energy. This is regardless whether the object is a piece of ice, a planet, a liquid, or a gas. The object radiates energy independent of its temperature. At low temperatures, most of the radiated energy is emitted in the infrared. At high temperatures, the objects emit in the visible, UV, and even in the X-ray part of the electromagnetic spectrum.

The stars emit electromagnetic radiation at different wavelengths, ranging from low infrared, visible to short UV wavelength, and in extraordinary circumstances, they emit X-ray and gamma radiation.

The intensity of the radiation emitted by the stars smoothly changes with wavelength. At low and high wavelengths, the intensity is low, and at intermediate wavelengths, the intensity is at its maximum, as shown in Figure 3-8.

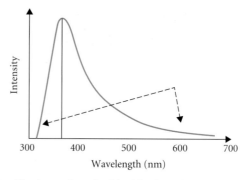

FIGURE 3-8. *The intensity of a blackbody radiator is low at short and long wavelengths.*

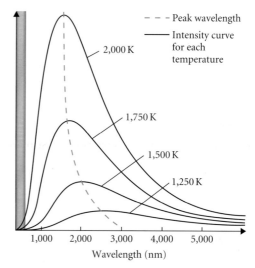

FIGURE 3-9. *The peak intensity and the temperature of a blackbody radiator are inversely related.*

The objects that emit continuous radiation are called **blackbody radiators**. The wavelength of the maximum intensity depends on the surface temperature of the object, as shown in Figure 3-9.

The continuous spectrum emitted by a blackbody radiator has the following characteristics:

1. The spectrum has a specific wavelength, or color, where the intensity of the radiation is maximum.

2. The intensity of the radiation is low at short and large wavelengths.

3. The wavelength of the maximum intensity depends exclusively on the temperature of the radiator.

An ideal blackbody radiator emits all the energy it produces and also absorbs all the energy it receives. In practice, only few ideal blackbodies are perfect absorbers and perfect emitters of radiation. In general, a blackbody looks black at low temperatures and looks bright at high temperatures. For example, at room temperature, charcoal looks black and at a moderately high temperature, it looks red.

In physics and astronomy, blackbodies are always associated with emitters and absorbers of energy. For example, the head light on our cars are blackbodies because they emit all the energy they

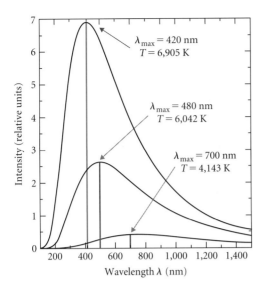

FIGURE 3-10. *The blackbody radiation of these stars tells us that the higher the temperature, the shorter the wavelength at which the intensity is maximum.*

produce. For the same reason, the stars and the Sun are considered blackbodies.

Two laws describe the behavior of blackbody radiators:

1. Wien's displacement law and
2. Stefan–Boltzmann law.

Wien's Displacement Law

The continuous radiation emitted by the surface of a blackbody smoothly changes with wavelength, and the wavelength at which the intensity is maximum shifts toward shorter wavelengths as the temperature increases. See Figures 3-9 and 3-10. This behavior is the essence of Wien's displacement law that has the following expression:

$$\lambda_{max} = \frac{2{,}900{,}000}{T} \qquad (3\text{-}9)$$

In this equation, the wavelength, λ_{max}, is expressed in nanometers (nm), and the temperature, **T**, in Kelvin.

Example

The blackbody radiation emitted by a star has its maximum intensity at a wavelength of 480 nm. What is the surface temperature of the star? See Figure 3-23.

Solution Solving equation 3-9 for **T** and replacing the numbers, we have

$$T = \frac{2{,}900{,}000}{\lambda_{max}} = \frac{2{,}900{,}000}{480} = 6{,}042 \text{ K}.$$

To measure the surface temperature of a star, we can follow the solution given in problem 11.

Q9. The Sun has a surface temperature of about 5,800 K. At what wavelength does the Sun emit the most?

a. 300 nm
b. 400.46 nm
c. 500 nm
d. 628.24 nm

Q10. According to Wein's law, the cooler a star, the redder it looks.

a. True
b. False

Q11. Which of the following needs to be done to measure the surface temperature of the Sun?

a. Take the continuous spectrum of the star
b. Determine the wavelength where the intensity of the radiation is at its maximum
c. Apply Wien's law
d. All the above

Q12. Tripling the temperature of a blackbody will triple the total energy it radiates.

a. True
b. False

Chapter 3—Interaction of Light with Matter

Q13. Star A has a surface temperature of 5,000 K and star B of 15,000 K. Both stars have the same radii. The energy emitted every second by one square meter of star B is _____ times the energy emitted by star A.

a. 6
b. 9
c. 27
d. 81

Q14. Figure 3-24 represents the continuous spectra emitted by four different stars. Which star has the highest surface temperature?

a. A
b. B
c. C
d. D

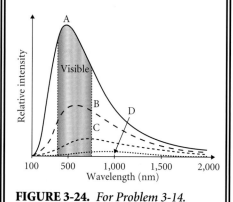

FIGURE 3-24. *For Problem 3-14.*

Stefan–Boltzmann Law

The energy emitted by a blackbody per square meter of surface per second is called **flux**. This flux is proportional to the temperature (T) raised to the fourth power:

$$F = \sigma T^4. \tag{3-10}$$

Since the flux is energy per square meter, the total energy (E) that a blackbody, such as a star, emits each second is equal to the flux times the surface area of the star or

$$E = \sigma A T^4. \tag{3-11}$$

In this equation, σ is a constant term. Ignoring the constant and assuming that the stars are spheres, the total energy they emit per second is proportional to the square of the radius times the temperature to the fourth power or

$$E \propto R^2 T^4. \tag{3-12}$$

This law is a form of the **Stefan–Boltzmann law**.

Example

Star A has a surface temperature of 5,000 K and star B has a surface temperature of 10,000 K, all other conditions being equal. How much more energy per second will star B radiate from its surface than star A?

Solution Since both stars have the same radii, the equation shown above indicates that the star with the highest temperature emits more energy. The ratio of the energies emitted by the stars is found by taking the ratio of the two temperatures and raising the result to the fourth power: (We can ignore the radii because they are identical.)

$$\frac{\text{Energy emitted by star A}}{\text{Energy emitted by star B}} = \left(\frac{10,000}{5,000}\right)^4 = 16$$

The energy emitted by star A is 16 times greater than the energy emitted by star B.

Atoms and Spectra

In addition to the continuous spectra emitted by blackbodies, there are two other types of spectra, the **absorption** and **emission line spectra**.

When the light from a blackbody passes through a relative cold gas, an absorption spectrum is produced. When a relatively low pressure gas is excited by photons or other means, it produces an emission line spectrum.

In order to understand the phenomena related to the production of spectra, we need to discuss atoms.

Atoms are the basic building blocks of matter. There are 90 different naturally occurring atoms. Approximately 25 more atoms have been artificially produced in laboratories.

Atoms are made of three basic particles: **protons, neutrons,** and **electrons**. The protons and neutrons are found in the nuclei of the atoms. The electrons whirl around the nucleus in an "electronic cloud", as shown in Figure 3-11.

Protons have positive charges, and neutrons have no charge. The mass of a neutron is a little larger than the mass of a proton. A proton has a mass of 1.67×10^{-27} kg. The mass of a proton is 1,840 times the mass of the electron.

The protons and neutrons are not fundamental particles; they are made of **quarks**. Electrons, however, are fundamental particles because they are not made of smaller particles. The protons and neutrons are called **baryons**.

It is believed that the number of protons in the universe is equal to the number of electrons.

The electrons in an atom are kept in place by the electrostatic force between the protons and the electrons. This force is proportional to $1/d^2$ just like gravity. The value of this force falls off quickly with growing distance.

The protons and neutrons are held together by the "strong nuclear force" that overcomes the electrostatic repulsion between the positive-charged protons.

The nucleus of the atoms contains protons and neutrons, except the hydrogen atom that has only a proton at the nucleus. Neutral atoms have as many protons as electrons, as shown in Figure 3-12.

"Although it is convenient to draw protons, neutrons, and electrons as little dots, quantum mechanics tells us that they cannot be located accurately and are in fact more like fuzzy little fog clouds. We cannot predict precisely what they will do, leading to a scientific confrontation with the philosophy of determinism: science shows that there is fundamental uncertainty in what will happen in the future." This quote is taken from http://ircamera.as.arizona.edu/NatSci102/NatSci102/lectures/spectroscopy.htm.

Neutral atoms can gain or lose electrons. This process forms **ions**. A positive ion is an atom that has lost one or more electrons and has an excess of positive charge as a result. A negative ion is an atom that has gained one or more electrons and as a result has an excess negative charge.

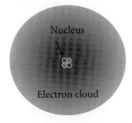

FIGURE 3-11. *The electron cloud swirls around the nucleus of the atoms.*

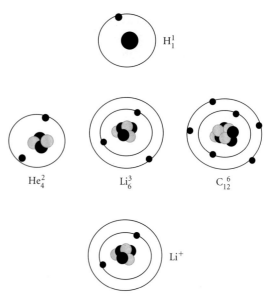

FIGURE 3-12. *Naïve representation of the neutral atoms of hydrogen (H), helium (He), lithium (Li), carbon, and the positive lithium ion (Li^+).*

FIGURE 3-13. *The water molecule has one oxygen and two hydrogen atoms.*

For example, a neutral lithium atom has three protons and three electrons. A lithium ion, by contrast, has lost one electron and has an excess of one positive charge. A lithium ion is expressed as Li^+. See Figure 3-12.

Two or more atoms bonded together form a molecule as shown in Figure 3-13.

In the following paragraphs, reference will be made mainly to the hydrogen atom because it is the most abundant atom in the universe. Further, what we learn about this atom can largely be applied to other atoms.

Electronic Energy Levels of the Hydrogen Atom

The electrons in an atom are found only in certain orbits, which are called **permitted orbits** or energy levels. Each atom has different orbits, or **electronic energy levels**, for its electrons.

The energy that attaches an electron to the atom is known as **binding energy**. Inner orbits have larger binding energy than outer ones.

Figure 3-14 shows the first four energy levels for the hydrogen atom. When the electron is in its lowest energy level, $n = 1$, the

Q15. What is the difference between a positive ion and a negative ion?

Q16. A positive hydrogen ion _____.

a. consists of a proton
b. has lost one electron
c. has gain one electron
d. has one proton and one electron
e. a and b

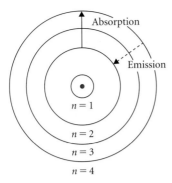

FIGURE 3-14. *Origin of the absorption and emission lines.*

atom is in the ground state. When the atom is in any other level, it is in an excited state.

A hydrogen atom becomes excited when its electron absorbs energy and is promoted to higher energy levels than 1, **n** = 2, 3 . . . etc.

If the electron absorbs too much energy, it will leave the atom and the hydrogen atom will become ionized.

In astronomy, we are particularly interested in studying the cases in which the electrons of the atoms absorb energy from photons.

Experiments have shown that the energy of the photon absorbed is equal to the energy difference between the energy levels where the transition of the electron occurs.

For example, if the incoming photon has an energy equal to the energy difference between orbit number 5 and 2, then the hydrogen atoms whose electrons are in orbit 2 will make an upward transition to orbit 5. The energy of the photon absorbed is equal to E5–E2. Other atoms whose electrons are in other orbits will not absorb these photons.

In general, when the electron of an atom absorbs energy, the electron makes an upward transition from a lower energy level to a higher energy level and becomes excited. However, in about 10^{-8} s, the electron falls to a lower orbit which liberates the energy it has gained. This process is called **emission line spectrum**.

If the energy of the incoming photon has enough energy, the electron will leave the hydrogen atom and become ionized. When hydrogen atoms absorb the UV radiation emitted by hot stars, it becomes ionized. Large regions of ionized hydrogen are called **HII regions**, and large regions of atomic neutral hydrogen are called **HI regions**.

Light Spectra

In general, the distribution of intensities of light versus wavelength is called **spectra**. In astronomy, we are most interested in studying the three different type of spectra mentioned earlier: **continuous**, **absorption**, and **line** or **emission spectra**.

Q17. When an electron of the hydrogen atom absorbs a photon, the electron _____.

a. drops to an orbit of lower energy
b. (jumps to an orbit of higher energy)
c. becomes a positive ion
d. becomes a negative ion

Q18. When the electron of a hydrogen atom "makes a downward transition," it _____.

a. absorbs a photon
b. (emits a photon)
c. becomes a negative ion
d. absorbs another electron

Continuous Spectrum

We know that blackbodies, like the stars, emit continuous radiation. The continuous spectrum is obtained when the light from a blackbody passes through a transparent prism, as shown in Figure 3-5. Remember that each color has its corresponding range of wavelengths. See Table 3-1. As previously explained, the wavelength where the radiation is maximum is given by Wien's law. The intensity of the spectrum shown in Figure 3-15 is maximum around 420 nm.

Absorption Spectrum

When the light from a blackbody passes through a cool, low-density diffuse gas, the electrons of the atoms of the gas absorb photons and make an upward transition to orbits of higher energy, exciting the atoms and, in some cases, ionizing them. The spectrum of the light that passes through a cool gas consists of the continuous spectrum with a series of dark lines on it, as shown in Figure 3-16. The dark lines are due to the energy that the electrons in the gas have taken away from the light in order to make the "transition to higher energy levels."

This type of spectrum is called **absorption spectrum**.

Hydrogen Absorption Lines

The number of dark lines and their wavelengths depend on the type of gas the light passes through.

For example, if the gas is hydrogen, the absorption spectrum will have four lines in the visible part of the spectrum (the **Balmer absorption lines**). In Figure 3-16, the name and the wavelength of each line are given.

The Hα absorption line located at 656.3 nm is produced when the electron in the hydrogen gas absorbs a 656.3 nm wavelength photon.

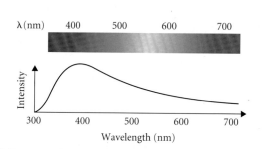

FIGURE 3-15. Blackbody radiators emit continuous radiation.

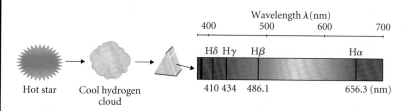

FIGURE 3-16. Visible absorption lines (Balmer lines) produced by hydrogen gas.

Q19. In the absorption spectrum of hydrogen, the Hα line (656.3 nm) has more energy than the Hδ line (410.2 nm).

a. True
b. False

Q20. The absorption spectrum of an element consists of dark lines superimposed on the continuous spectrum.

a. True
b. False

In this case electron makes an upward transition from energy level 2 to energy level 3. The Hβ absorption line located at 486.1 nm is produced when the electron in the hydrogen gas absorbs a photon of 486.1 nm of wavelength and jumps from energy level 2 to 4. The Hγ line located at 434 nm is formed when the absorbed photon promotes the electron from orbit 2 to orbit 5. The Hδ line located at 410.2 nm is produced when the electron goes from orbit 2 to orbit 6. The Hα line is called the **red line** because it is in the red part of the spectrum. The Hβ line is the blue line, Hγ is the blue-violet, and the Hδ line is located in the violet.

The continuous spectra of the stars always show the dark absorption lines of the elements they bear. These lines are produced when the photons coming from the photosphere (surface) are absorbed in the cooler regions of the lower atmosphere of the stars.

Comparing the spectra of the stars with the spectra of the known elements, the chemical compositions of the stars, planets, Sun, and interstellar gases must be obtained. Figure 3-17 shows the absorption lines of the Sun in the visible part of the spectrum.

> **Q21. The absorption spectrum of the a star gives information about the star's _____.**
> a. size
> b. name
> c.) composition
> d. location
>
> **Q22. The _____ spectra of the Sun and stars are produced when the light coming from their photospheres passes through the cooler gases of their atmospheres.**
> a.) absorption
> b. an emission line
>
> **Q23. What is the difference between an absorption and a continuous spectra?**

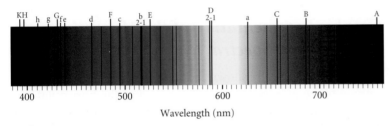

FIGURE 3-17. *Sun's Visible absorption spectrum lines.*

Emission Line Spectrum

We saw that an atom becomes excited when one of its electrons absorbs a photon. However, excited atoms are unstable, and in a few nanoseconds, the electron makes a downward transition emitting a photon of energy equal to the energy it absorbed. This process produces an **emission line spectrum** or **line spectrum**. See Figure 3-18.

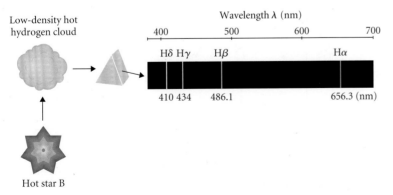

FIGURE 3-18. *The hydrogen cloud absorbs energy, and the electrons become excited. As they return downward, they emit the absorbed energy which produces a line spectrum.*

Chapter 3—*Interaction of Light with Matter*

If the gas is hydrogen, the spectrum has four lines. The wavelength of each emission line corresponds to the wavelength of the absorption wavelengths, as shown in Figure 3-19. The dark background in the emission spectrum is due to the lack of emission lines in these regions.

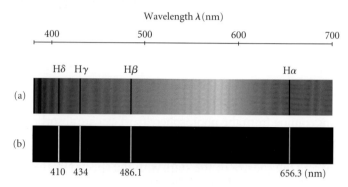

FIGURE 3-19. (a) *Absorption lines of hydrogen.* (b) *Emission lines of hydrogen.*

Graphic Representation of the Absorption and Emission Spectra

The spectra are usually shown in a graph, where the vertical axis represents the intensity of the radiation, and the horizontal axis represents the wavelength of the absorption or emission lines. In the absorption spectrum, the absorption lines appear as dips or valleys. The dips represent the wavelengths in which the atoms in the gas absorb energy. See Figure 3-20.

In the emission spectra, the lines appear as discrete positive lines, located at wavelengths where the energy is emitted. See Figure 3-21.

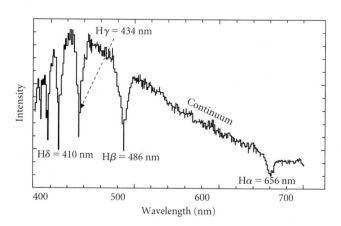

FIGURE 3-20. *Absorption spectrum of hydrogen.*

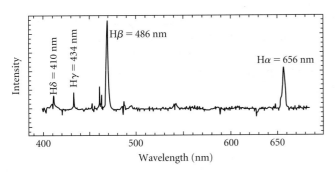

FIGURE 3-21. *Hydrogen emission line spectrum.*

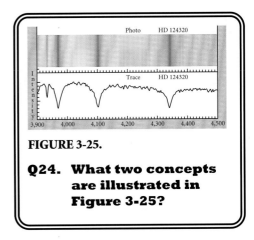

FIGURE 3-25.

Q24. What two concepts are illustrated in Figure 3-25?

Emission Nebula

The Milky Way is a relative young galaxy with millions of stars still forming. In later chapters, we will learn that the stars form inside huge hydrogen clouds. Therefore, many hot stars are still surrounded by clouds of atomic hydrogen.

These hot stars emit large amounts of UV radiation that ionizes the nearby hydrogen atoms and produces large numbers of free electrons. The free electrons move with random motion inside the ionized hydrogen cloud. When the free electrons get near the positive hydrogen ions, they are captured. In astronomy, we say that the electrons "recombined" with the hydrogen ions. The recombination process emits the photons corresponding to the four Balmer emission lines: the red (the Hα), the blue (Hβ), the blue-violet (Hγ), and violet (Hδ). The mixture of these four photons has a characteristic red-pink color. The red color dominates because the intensity of the red Hα line is stronger than the intensity of the other lines. Therefore, the hydrogen clouds that contain hot stars glow with a red-pink color. These regions are called **emission nebulae**.

There are millions of emission nebula in our galaxy and in other galaxies.

Emission nebulae are called HII regions because they are regions of ionized hydrogen. Regions of neutral hydrogen are called HI.

The Doppler Effect

The Doppler effect is a change in the wavelength or frequency of a wave due to the motion of the source of the wave and/or the observer. The effect is most easily visualized with sound waves propagating on air.

Consider a police car with the siren wailing while waiting for the red traffic light to change to green. Consider also two observers—one standing in front of the car and the other in the back of the car—at a certain distance **d**, as shown in Figure 3-22a. The two observers will hear the siren with the same frequency and with the same wavelength.

Q25. By analyzing starlight, astronomers are able to determine the star's _____.

a. radial velocity
b. composition
c. temperature
d. all the above

Q26. The red absorption line Hα of a star was measured to be 658.1 nm. This information tells that the star is moving _____ the observer. (Hint: look for the value of the unshifted line.)

a. toward
b. away from

Q27. The Hβ absorption line from a star was measured to be 485.1 nm. This information tells that the line is _____ shifted. (Hint: Look for the value of the unshifted Hβ line.)

a. blue
b. red

Q28. The Hβ absorption line from a star was measured to be 485.1 nm. What is the value of $\Delta\lambda$ in nanometer?

a. 10
b. 1.0
c. -1.0
d. 0.45

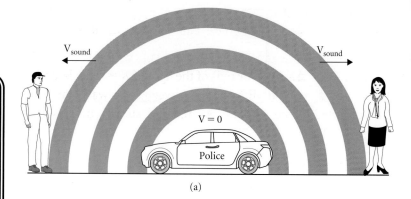

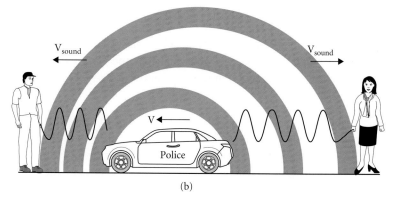

FIGURE 3-22. (a) *The two observers hear the same frequency.* (b) *The observer to the left hears a higher frequency than the one to the right.*

Now, consider what happens when the car moves toward the observer on the left, with constant speed **V**. This observer will see that the wave crests arrive closer together than when the car was standing still. Therefore, he hears the sound with higher frequency or smaller wavelengths. The observer on the right, behind the car, sees the crests of the waves farther apart because, to the speed of the sound you have to subtract the speed of the car. Therefore, this observer hears the sound with lower frequency, i.e., longer wavelength. See Figure 3-22b. The apparent change in the wavelength of sound caused by the motion of the source (or of the observer) is called **Doppler effect**.

With light, the Doppler effect is particularly observed in the absorption spectra. If a star is not moving with respect to the Earth, the wavelengths of the absorption spectrum are in the right place. This wavelength is called the **rest wavelength λ_0**. See Figure 3-23a.

If the star is moving away from the observer, as seen in Figure 3-23b, the wavelength of the absorption spectrum will be shifted to longer wavelengths, i.e., toward the red (the spectrum will show a red shift).

Conversely, if the star is moving toward the observer, the absorption spectrum is shifted toward the blue part of the spectrum. In this case, the spectrum is blue shifted, as shown in Figure 3-23c.

The absorption spectrum of the stars informs us whether they are moving away from or toward us.

The Doppler effect is only sensitive to the part of the velocity directed toward or away from us. This is called **radial velocity**.

The radial velocity of the stars can be obtained by measuring the amount of shift $\Delta\lambda$ (blue or red) of the light [$\Delta\lambda$ = delta lambda]. $\Delta\lambda$ is equal to the difference between the wavelength of the unshifted spectrum and λ_0 and the shifted or observed wavelength λ

$$\Delta\lambda = \lambda_{observed} - \lambda_0.$$

The radial velocity of a star that shows a shift in the absorption spectrum is found using

$$V_{radial} = \frac{\lambda_{observed} - \lambda_0}{\lambda_0} c = \frac{\Delta\lambda}{\lambda_0} c,$$

where **c** is the speed of light and is equal to 300,000 km/s.

When the stars move away from us, $\Delta\lambda$ is positive.

When the stars move toward us, $\Delta\lambda$ is negative. Why?

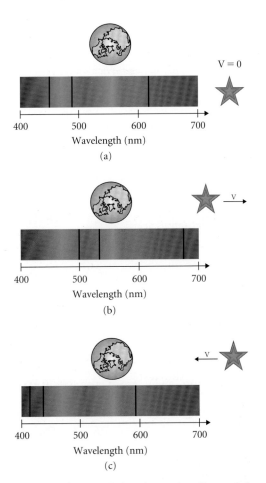

FIGURE 3-23. *The position of the absorption lines of the absorption spectrum depends on the relative motion of the stars (a) the star is at rest wit respect to Earth, (b) the star is moving away from Earth and (c) the star is moving toward Earth.*

The different wavelengths of the absorption lines of hydrogen are shifted by a different amount because the shift is proportional to each unshifted wavelength λ_0.

If we take the absorption spectrum at one edge of a star and then at the other end, and find that one spectrum is red shifted and the other is blue shifted, we conclude that the star is rotating on its axis. Why?

Astronomers use the Doppler effect quite frequently.

ANSWERS

1. c, yes, these stars exist and rotate on their axis extremely fast, which means that they have a small period.
2. c
3. d
4. b
5. c
6. b
7. b
8. c
9. c
10. a, The cooler a star, the longer the wavelength where the intensity of the radiation is maximum. Red color has longer wavelength than other color.
11. d
12. b, The energy will be 81 times larger
13. d, $[15,000/5,000]^4 = 3^4 = 81$
14. a, The spectra that has the maximum at the shortest wavelength represents the star with the largest surface temperature.
15. A positive ion has more protons than electrons, and a negative ion has more electrons than protons.
16. e
17. b
18. b
19. b
20. a
21. c
22. a
23. The continuous spectrum is emitted by a blackbody and its intensity smoothly varies with wavelength. The absorption spectra are produced by cool gases when the light from a blackbody passes through them. The intensity of the absorption spectrum is discrete, forming dark lines on the continuum spectrum.
24. That the absorption spectrum of an element can be presented by means of a photograph, top diagram, or diagram displaying wavelength versus intensity.
25. d
26. b, The wavelength of this Hα line is larger than unshifted Hα (656.3 nm) line.
27. a
28. c, $485.1 - 486.1 = -1.0$

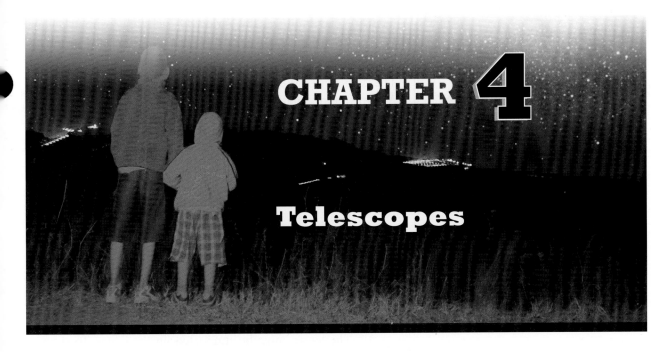

CHAPTER 4

Telescopes

The electromagnetic radiation emitted by the stars, galaxies, and other celestial objects travels through space. This radiation ultimately reaches Earth where it can be detected mainly with telescopes. Each type of radiation requires a different detector. For example, optical telescopes detect only the visible part of the electromagnetic spectrum. Observations in other regions require other types of telescopes. Astronomers routinely use radio, infrared, and ultraviolet telescopes. The space age has extended the observations to include the X-ray and gamma ray regions. In this chapter, we are going to briefly study the main types of telescopes used by astronomers.

OPTICAL TELESCOPES

Optical telescopes are used to collect visible light and bring that light into focus.

There are two types of optical telescopes: **refractors** and **reflectors**.

Refractors

Refracting telescopes or refractors, have a big lens (objective) that gathers the light and directs it to a focal point. A small lens (eyepiece) brings the image to your eye, or detector, as indicated in Figure 4-1. A film or a charge-couple device (CCD) camera is placed at the focal point where the image forms. The largest refractor

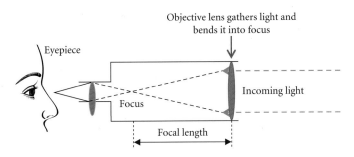

FIGURE 4-1. *Main elements of a refractor.*

> **Q1.** Which of the following telescopes can be used to make observations from the surface of Earth? (Hint: see Figure 3-6).
>
> a. X-ray telescope
> b. gamma ray telescope
> c. far infrared telescope
> d. RT

telescope ever built was the **Yerkes 40-inch** (approximately equal to 1.02 m). When it was completed (in 1897), it was the largest telescope in the world.

The diameter of the objective indicates the size of the telescope. A 2-m (approximately 6 feet) telescope has an objective lens of 2-m of diameter.

Refractors have drawbacks that limit their size and applications. The main complications with refractors are as follows:

1. Each color has a different index of refraction which means that the color bends differently when it passes through the lens of the telescope. This causes the colors to disperse around the image. See Figure 4-2. In astronomy, this defect hinders the quality of scientific observations.

2. The lenses are so large that they are rarely free of defects. Further, it is difficult to make a lens with perfect curvature on *both sides*.

3. The lenses must be supported on the edges. To support heavy lenses, which are bulky and heavy, large structures are needed for support. Such large structures are difficult to control and to guide.

4. The glass of the refractors absorb the infrared and the ultraviolet (UV) radiation in the light. Therefore, refractors cannot be used to observe light in the these wavelengths.

For these reasons, large optical telescopes are not refractors. All large telescope telescopes are reflectors.

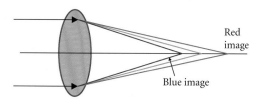

FIGURE 4-2. *Each color has a different index of refraction. Therefore, each color has a slightly different focus.*

Reflecting Telescopes

Reflecting telescopes, or **reflectors**, use a curved mirror (primary mirror) to bring the light into focus. See Figure 4-3. In contrast to refractors, reflectors do not present chromatic aberration. Further, they do not absorb any portion of the visible light because the light does not pass through the glass. Reflectors are lighter than refractors, they have only one surface to polish, and they are supported in the back.

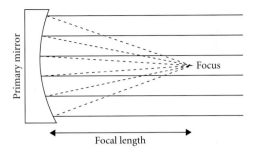

FIGURE 4-3. *The light is brought into focus by means of a lens.*

Two Important Type of Reflectors

The Newtonian and the Cassegrian reflectors are two common types of reflecting telescopes. The Newtonian design is very popular for small telescopes. This type of telescope has a secondary mirror that deflects the image out of the telescope to the eyepiece. The image forms at this point instead of forming at the focal point **F'** inside the telescope. See Figure 4-4.

The Cassegrian telescope uses a secondary mirror to deflect the image to the rear of the telescope where the instruments and detectors are located. With some modifications, this type of design is very popular in large telescopes.

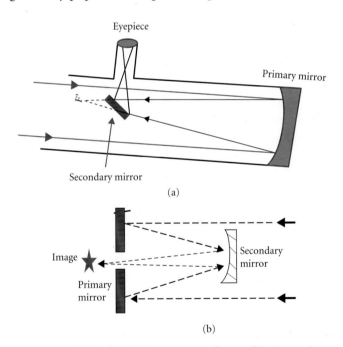

FIGURE 4-4. (a) *Sketch of a Newtonian reflector.* (b) *Cassegrian.*

LARGE OPTICAL TELESCOPES

At present, the largest operating optical telescope in the world is the 10-m twin Keck reflector at Mauna Kea, Hawaii, 13,796 feet (4,205 m) above sea level.

> **Q2. The atmosphere of Earth is totally opaque to what kind of radiation?**
> a. UV
> b. infrared
> c. radio
> d. gamma rays

The primary mirror of this telescope is so large that it was impossible to make it with a single piece of mirror. Instead, the telescope consists of an array of 36 hexagonal mirrors, each mirror having a diameter of 1.8 m (~ 6.0 feet).

The focal length of the array is 17.5 m (~ 57.4 feet). The entire structure is computer-controlled, and the 36 mirrors work as a collective single unit.

The high altitude of the observatory and the large primary mirror of the twin telescopes make them particularly useful to study faint objects in the near infrared and in the visible part of the spectrum.

The twin Keck telescopes are the world's largest optical and infrared telescopes. Each telescope has the Cassegrian geometry and they work separately and together to form a single interferometer.

Later in the chapter, the concept of interferometer will be explained.

Mount Graham, in southeastern Arizona, is the seat of the "Large Binocular Telescope (LBT)." It consists of two 8.4-m reflectors mounted on the same structure to increase its ability to gather light. In the future, it will operate in the interferometer mode. The primary mirror of each of these two telescopes has a diameter of 11.8 m (39 foot).

The LTB is the largest single mirror telescope in the world and operates in the UV, visible, and infrared part of the spectrum. For more information visit the Web site http://medusa.as.arizona.edu/lbto/.

In Cerro Paranal, Chile, there are four reflectors each with a single primary mirror of 8.2 m in diameter. These telescopes work individually or link together forming an interferometer and have the capability to make observations in the visible and in the infrared parts. These telescopes are known as the Very Large Telescope Interferometer (VLTI). For more information, visit the Web site: http://www.eso.org/projects/vlare/.

In the Atacama desert in Chile, the Giant Magellan Telescope (GMT) is under construction.

Scheduled for completion around 2017, the GMT will have the resolving power of a 24.5-m (80 foot) telescope. When commissioned, this telescope will be the largest optical telescope ever constructed, and it will answer many of the questions at the forefront of astrophysics today. For more information, visit the Web site http://www.gmto.org/.

Table 4-1 gives a summary of the largest reflectors in the world.

SPECIAL INSTRUMENTS IN TELESCOPES

Telescopes need instruments to record, analyze, and store the radiation collected from outer space.

TABLE 4-1. *The largest optical reflectors in the world.*

NAME	APERTURE	TYPE	SITE
Large Binocular Telescope (LBT)	8.4 m × 2	Single	Mount Graham
Southern African Large Telescope (SALT)*	11.0 m	Mosaic	South Africa
Gran Telescopio Canarias (GTC)*	10.4 m	Mosaic (IR)	Canary Islands
2- Keck	10.0 m	Mosaic	Mauna Kea
Subaru	8.3 m	Single (IR)	Mauna Kea
4- Very Large Telescope (VLT)	8.2 m	Single	Paranal, Chile
Gemini 1 North	8.1	Single	Mauna Kea
Gemini 2 South	8.1	Single	Chile

* In commission, and not fully operational yet.

The images of most modern telescopes are recorded and stored by means of **electronic imaging systems**, such as a CCD. CCDs are arrays of light-sensitive diodes (detectors), similar to the detectors used in home video cameras. Most telescopes store the images for analysis directly into a computer. The light gathered by the telescopes from the stars and other celestial entities is analyzed by means of an instrument called the spectrograph. A spectrograph spreads out the light according to the wavelength to generate the light spectra. The spectra reveal the information that the light contains.

The spectrograph is one of the most important instruments of a modern large telescope.

The telescopes must have a **sidereal drive**, which compensates for the Earth's rotation, as shown in Figure 4-5. Without a sidereal drive, a telescope loses its target quickly.

POWERS OF A TELESCOPE

The diameter of the primary mirror of a telescope determines both the **light gathering power** (LGP) and the **resolving power** of a telescope.

Q3. Large optical telescopes are always of the _____ type.
a. refractor
b. reflector
c. either reflector or refractor

Q4. Chromatic aberration is a problem with _____.
a. refractors
b. reflectors

Q5. The largest optical telescope in the world is (are) the _____, located in _____.
a. Keck; Hawaii
b. Gemini; Germany
c. LBT; Arizona
d. Palomar; California

Q6. Two drawbacks of refractor telescopes are:

Q7. The primary mirrors of the twin Keck telescopes consist of an array of _____ mirrors.
a. 20 circular
b. 32 pentagonal
c. 36 hexagonal
d. square

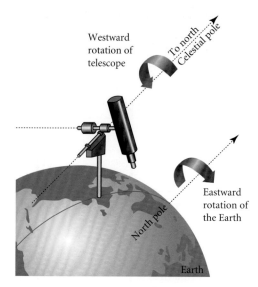

FIGURE 4-5. *The sidereal drive compensates for the rotation of the Earth. Therefore, the telescope can look at the same object for several hours.*

Light Gathering Power

The LGP is probably the most important feature in a telescope. It describes the amount of light, or photons, that pass through the primary mirror. The larger the diameter, **D**, the greater the LGP and the greater the chance to see faint objects. One of the reasons that astronomers use large telescopes is to increase the LGP of the telescopes.

The LGP of a telescope is proportional to the surface area or the square of the diameter (D^2) of the primary mirror. This is written in an equation as

$$\text{LGP} \propto D^2 \qquad (4\text{-}1)$$

Since the diameter is equal to two radii, (**D = 2R**), we can write

$$\text{LGP} \propto R^2$$

The number 4 is dropped in this last equation because we are interested in proportions only. An example will help to illustrate the importance of this relation.

Example

A 1.5-m telescope produces an image four times brighter than a 0.75-m telescope, because $\left[\dfrac{1.5}{0.75}\right]^2 = 2^2 = 4$. Similarly a 4-m telescope gives an image about 10 times brighter than a 1.25-m telescope because $\left[\dfrac{4}{1.25}\right]^2 = 10.24$.

The LGP of the 10-m Keck telescope has four times the LGP of the 5-m Palomar telescope. Why?

> **Q8. If the diameter of the primary mirror of a telescope is tripled, its LGP will increase by a factor of _____.**
>
> a. three
> b. one-third
> c. nine
> d. six
>
> **Q9. Telescope A has a diameter of 0.25 m and telescope B has a diameter of 0.5 m. The LGP of telescope A is _____ times the LGP of telescope B.**
>
> a. 1/4
> b. 4
> c. 2.25
> d. 1/5

Angular Resolution or Resolving Power

Angular resolution, or **resolving power**, refers to the ability of our eyes, a camera, or a telescope to show fine detail and to differentiate images of objects that are separated by small angles.

Our eyes are able to resolve objects or features that have an angular separation larger than 0.5 arc minute.

The minimum angular separation that a detector, telescope, or camera can distinguish is a measure of the resolving power of that instrument. The stars are far away and have small angular diameter. Therefore, with even the most powerful telescopes, we cannot see any surface details of the stars. New techniques are offering up new possibilities to observe the stars. In 1995, the Hubble Space Telescope (HST) was able to image the red giant star, Betelgeuse, using its improved near-UV camera.

Another example will clarify the concept of angular resolution.

Example

Consider the two stars separated by an angle of 0.44 arc second. When the two stars are viewed by a telescope with an angular resolution of 0.55 arc second, we get a blurred image. However, when using a telescope that resolves angles of less than 0.44 arc second, the stars appear as two different points.

The resolution of a telescope is limited by a physical property of the lenses and mirrors called **diffraction**. Diffraction is the bending of waves when they encounter a barrier or pass through an aperture. Diffraction produces small fringes around every point of the image, making the image fuzzy. We cannot see any detail smaller than the diffraction fringes. When the diffraction fringes of two stars overlap each other, the images cannot be resolved.

The diffraction effect cannot be eliminated. However, it can be minimized by making the diameter of the primary mirror as large as possible. In general, large primary mirrors produce smaller fringes and better resolution.

The diffraction also depends on the wavelength of the radiation in which the observations are made.

The angular resolution, or resolving power (α), of a telescope depends on both the diameter of the telescope and on the wavelength of the radiation being used. The smallest angle that diffraction allows, is given by

$$\alpha \text{ (arc second)} = 0.25 \frac{\text{Wavelength } \lambda \text{ (nm)}}{\text{diameter of primary D (mm)}} \quad (4\text{-}2)$$

This is known as the **diffraction limit**.

Notice that in equation 4-2, the wavelength is given in nanometers (nm) and the diameter in millimeters (mm).

Q10. What is the resolving power of a 5-m telescope that uses near infrared light of 1,000 nm of wavelength? Hint: Do not forget to convert meters (m) to nanometers (nm).

   a. 0.125 arc second

   b. 0.50 arc second

   c. 0.2 arc second

   d. 0.05 arc second

Q11. Telescope A has a diameter of 40 cm and telescope B has a diameter of 20 cm. If both telescopes are using the same type of radiation, how much more powerful is telescope A than B? Hint: Since both telescopes are using the same wavelength, take the ratio of the diameters, keeping in mind that the larger diameter, the smaller the angle the telescopes resolve and the more powerful they are.

   a. two times

   b. one-half

   c. four times

   d. one-fourth

We need telescopes that resolve small angles. According to equation 4-2, this is achieved when the telescopes have large diameters and use short wavelengths.

The 200-inch (5 m) Palomar Observatory has twice the resolving power of the 100-inch Mount Wilson Observatory.

Equation 4-2 also indicates that the shorter the wavelength of the radiation used, the smaller the angle that a telescope resolves. Thus, visible wavelength gives better resolution than infrared light.

Example

What is the minimum angle that a 1 m (1,000 mm) telescope, that uses a wavelength of 520 nm yellow light, can resolve?

Using equation 4-2, we arrive at the following solution:

$$\alpha = 0.25 \frac{\lambda \, (nm)}{D \, (mm)} = 0.25 \frac{520}{1,000} \text{ arc second}$$

$$= 0.13" \text{ arc second}$$

The minimum angle that diffraction allows for this particular telescope is 0.13 arc second; i.e., if two stars in the sky are separated by 0.13 arc second or more, the telescope would resolve the two stars.

In principle, this is the smaller angle that this telescope can resolve. In practice, the Earth's atmosphere degrades the resolving power of the telescope, and the telescope can only resolve objects that are separated in the sky by 0.65 arc seconds, instead of by 0.13 arc seconds.

If the same telescope were using infrared light of 1,000 nm, the resolution would be only 0.25 arc second.

Example

What is the minimum angle that a 2.4-m telescope, that uses visible light of 520 nm, can resolve?

Solution Using equation 3-2, we find

$$\alpha = 0.25 \frac{\lambda \, (nm)}{D \, (mm)} = 0.25 \frac{580}{2,400} \text{ arc second}$$

$$= 0.06" \text{ arc second}$$

0.06 arc second is the smallest angle that diffraction allows a 2.4-m telescope to resolve. The conditions of the Earth's atmosphere (called the seeing condition) lowers the resolving power of the telescope to about 0.5 seconds of arc.

The HST has a diameter of 2.4 m and has a resolution close to 0.06 arc second because it is above the Earth's atmosphere.

The Seeing Conditions of the Atmosphere and Adaptive Optics

The ultimate angle that a telescope resolves is determined by the Earth's atmosphere, regardless of the size of the diameter of the telescope. The atmospheric conditions of weather and air turbulence, called the seeing conditions, considerably lowers the theoretical angular resolution of the optical telescopes predicted by diffraction.

For example, equation 4-2 states that a 10-m telescope, using yellow light of 520 nm of wavelength, will resolve features that are separated by 0.013 arc seconds, but the atmosphere conditions increase this angle to about 0.13 arc second.

The light from the stars goes through miles of turbulent air in the Earth's atmosphere, causing the images from the stars to twinkle. The twinkling and the blurring of the image is larger when the stars are in the horizon, where we look through more air.

Optical telescopes perform better on high mountain tops where the atmosphere is thinner. They perform even better when they are in orbit above the Earth's atmosphere.

The best sites in the world to position optical telescopes are Manua Kea, Hawaii, and the Atacama desert in Chile. Besides being located at high altitudes, the atmospheres are dry and stable. In the United States, the best places for optical telescopes are in Arizona and New Mexico.

In Figure 4-6, you can see the difference when the same object is observed through the Earth's atmosphere with a ground-based telescope and without the atmosphere with the HST.

The use of a technique called **adaptive optics** (ADO) greatly improves the images taken with Earth-based telescopes. This is usually done with a reference star in the telescope's field of view or with an artificial guided star generated by a laser light. High-speed computers adjust the telescope optics, deforming the surface of the mirror (objective) to compensate for the atmospheric changes.

Using ADO, Earth-bound telescopes give a resolution close to the diffraction limits predicted by equation 4-2.

Many large telescopes use this technique, at least for part of their observations, producing images comparable in resolution to the ones obtained with the HST.

Using ADO, astronomers are able to compensate for the minor changes of the structure of the telescopes due to the fluctuations in temperature during the time of the observation.

Q12. If we halved the diameter of a telescope, the resolving power goes down by a factor of 1/2.

a. True, because the angle that it resolves is twice as large

b. False, because the angle that it resolves is reduced by two

Q13. The angle that a 12-inch diameter telescope resolves is _____ times smaller than that of a 4-inch diameter telescope.

a. 3
b. 4
c. 16
d. 9
e. 81

Q14. The 400-inch Keck reflector can see objects _____ times fainter than the 40-inch Yerkes lens.

a. 100
b. 10
c. 16
d. 160

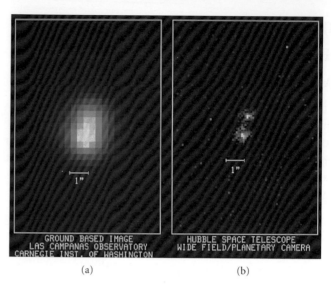

FIGURE 4-6. *Effect of the atmosphere on the images of telescopes.* (a) *Ground base image, Las Campanas Observatory.* (b) *Hubble Space Telescope.*

Credit: NASA Hubble Space Telescope Collection

Conclusion

To minimize the effect of atmospheric conditions on observations made by optical telescopes, astronomers

1. locate the telescopes on mountain tops and
2. use adaptive and active optics during observations.

However, the best results are obtained when the telescopes are located in space.

OPTICAL INTERFEROMETRY

The resolving power of telescopes is significantly improved when the images of two or more telescopes are combined using a technique called **interferometry**. For this purpose, the telescopes are linked with fiber optics to bring the information to a common unit where the signals are mixed. The signals from the different telescopes have to arrive in phase, to the mixing center, to produce constructive interference. If the signals are out of phase, destructive interference occurs and the image obtained is of poor quality.

By linking the telescopes, the effective area of the telescopes is "electronically" enhanced, increasing the resolution of the system. The LGP is not affected.

Optical interferometry is currently used in several observatories around the world. One of the most modern observatories is located in Cerro Paranal in northern Chile. It consists of four large telescopes known as the Very Large Telescope Interferometer (VLTI). When the telescopes are used in the interferometer mode, they provide the resolving capability of a single 200-m telescope.

The two 8.4-m telescopes of the VLTI will also operate in the interferometric mode to provide a resolution of a 22.8-m telescope. (The selection of the site for the VLTI sparked opposition from the local Apache tribe and environmentalists.)

The Center for High Angular Resolution Astronomy (CHARA) is operated by the Georgia State University on Mount Wilson. This observatory is one of the largest optical/infrared systems ever built. The system consists of six 1-meter telescopes that are linked together to form an **optical/infrared interferometer**. The six telescopes are dispersed over the mountain to provide a two-dimensional layout that provides the resolving capability of a single telescope a fifth of a mile in diameter.

This array is capable of resolving details as small as 0.2 arc seconds, equivalent to the angular size of a nickel seen from a distance of 10,000 miles. The Web site of the observatory is http://www.chara.gsu.edu/CHARA/.

The Twin Keck telescopes also can be used in the interferometric mode.

It is important to realize that interferometry only improves the angular resolution of the telescopes. The LGP or sensibility remains intact.

Observatories

The sites of observatories are carefully selected. They are located far away from cities to avoid light pollution. They are also located on top of high and dry mountains, which minimizes the negative effect of atmosphere conditions.

Dry air is very important for observations in the near infrared. The driest place on Earth is the Atacama desert, where several observatories have been built. Another excellent site is Mauna Kea, Hawaii. This is the home of the Twin Keck telescope and of the infrared Subaru telescope.

Within the continental United States, Arizona and New Mexico offer good sites to place optical telescopes. The Kitt Peak National Observatory (KPNO) is located on Kitt Peak in the Quinlan Mountains 90 km southwest of Tucson, Arizona. However, light pollution is becoming a problem in this observatory.

Radio Telescopes

Telescopes receive and record the **radio radiation** coming from outer space.

Q15. The instability of the atmosphere causes the stars to twinkle.
 a. True   b. False

Q16. In the future, ADO will greatly enhance the resolution of the HST.
 a. True   b. False

Q17. The use of ADO improves the images taken with the telescopes because:

Q18. The resolution of a telescope depends upon:
 a. the wavelength used and the diameter of the telescope's objective lens or mirror
 b. the design of the telescope
 c. whether the telescope is a reflector or refractor
 d. the brightness of the object
 e. the size and sensitivity of the CCD chip used for imaging

Q19. What problem does ADO correct?
 a. defects in the optics of the telescope, such as the original Hubble mirror
 b. the opacity of the Earth's atmosphere to some wavelengths of light
 c. the light pollution of urban areas
 d. turbulence in the Earth's atmosphere
 e. chromatic aberration due to use of only a single lens objective

Q20. An 18-cm telescope can be bought for a few hundred dollars. Galileo's telescope was about 6 cm. How much more light does the 18-cm telescope gather than Galileo's telescope gathered?

a. 1/9
b. 1/3
c. the same amount
d. 3
e. 9

Q21. The Chara system of six 1-m telescopes, located on Mount Wilson, _____.

a. can be used as an optical/infrared interferometer
b. when used as an interferometer, the resolution provided far exceeds the resolution of one individual telescope
c. when used as an interferometer the light gathering provided far exceeds the resolution of one individual telescope
d. all of the above
e. a and b

Q22. A 5-inch telescope using red light will provide better angular resolution than when used with blue light.

a. True
b. False

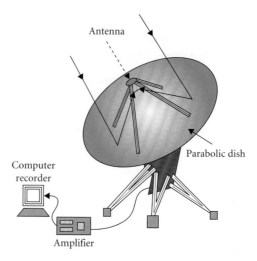

FIGURE 4-7. *Main elements of a radio telescope.*

The main parts of a radio telescope (RT) are as follows:
1. the dish (reflector),
2. the antenna at prime focus, and
3. the amplifier and record.

See Figure 4-7.

The reflecting surface of the dish is usually large and does not have to be as smooth as the surface of the primary mirror of the optical telescopes.

The radio radiation that the dish reflects is collected by an antenna. From the antenna, the signal travels through a cable to an amplifier and then to a computer, where it is stored and analyzed.

Two outstanding RT are
1. the 100-m RT in Green Bank, West Virginia, and
2. the 1,000 feet RT in Arecibo, Puerto.

The 100-m RT in Green Bank, is the largest steerable telescope in the world.

The images of RTs are presented in contour maps with false color. For comparison purposes, the optical image of the same galaxy is also displayed.

Resolving Power and Interferometry of RTs

RTs use long wavelengths and therefore, they have poor angular resolution. For a RT to provide a decent resolution, they must be very large. Astronomers overcome the low angular resolution of RTs by using them in the interferometer mode.

As we saw earlier, interferometers combine information from several widely spread RTs to provide a resolving power equivalent to the resolution of a RTs whose diameter is equal to the separation of the telescopes.

Remember that interferometry requires the signal from the different telescopes of the array to arrive in phase to the mixer of the signals. If the signal is out of phase, the image obtained is not of good quality.

A good example of radio interferometers is the Very Large Array (VLA) interferometer located in the New Mexico-Arizona desert. It consists of 27 telescopes connected to each other. The array gives a resolution of approximately 1 arc second, but the LGP of the interferometer remains equal to that of a single telescope.

Eight new RT are being added to the Very Large Array. This addition will increase the resolution to about 0.1 arc second.

Another example of a radio interferometer is the **Very Long Baseline Array**, which is a system of 10 radio telescope antennae, each with a dish of 25 m (82 feet) in diameter. These antennae are located from Hawaii to the Virgin Islands. When finished, it will be equivalent to the area of the country. The observations in this array will not be made simultaneously. The telescopes will observe the same objects at different times and record the results of their observations. Later, the data will be brought together and analyzed with computers.

RTs have large dishes (reflectors) not only to improve the resolution but also to collect as much radiation as possible since radio signals from space are very weak. These telescopes need to be located in remote places and in low valleys to avoid interference from terrestrial equipment.

Despite the poor angular resolution of RTs, they are important for the following reasons:

1. They are the only instruments able to detect cool interstellar clouds of hydrogen gas.
2. They can be used 24 h a day. Most RTs can be used during rain or snow storms because the radio waves are longer than the average rain drops or snow flakes.
3. Many celestial objects are poor emitters of visible radiation but are strong radio emitters. The center of the Milky Way is one good example.
4. They are effective because radio radiation is unaffected by interstellar dust, which absorbs visible radiation.

RTs are the only instruments capable of detecting the 21-cm wavelength emitted by the clouds of cold atomic hydrogen that fill the Milky Way and other galaxies. This is a very important feature since approximately 90% of the atoms of the universe are made of hydrogen.

Near Infrared Telescopes

As mentioned earlier, the atmosphere is transparent to the near infrared but opaque to the far infrared radiation. Therefore, the first group of telescopes can be located on Earth. However, the second group of telescopes must be located in space. (Near infrared has a wavelength of between 700 to approximately 2,000 nm. Far

---

Q23. Which power of a telescope might be expressed as "0.08 arc seconds"?
   a. LGP
   b. Resolving power or angular resolution
   c. Magnifying power
   d. Both a and b above

Q24. Two or more telescopes are linked to form an interferometer to
   a. increase angular resolution
   b. increase the LGP
   c. be able to image fainter objects

Q25. It is easier to set up RTs as an interferometer than optical telescopes because the radio radiation has longer wavelengths than the visible radiation.
   a. True
   b. False

Q26. A RT has a better resolving power than an optical telescope because the radio radiation has longer wavelengths than the visible radiation.
   a. True
   b. False

infrared has a longer wavelength than 2,000 nm.) The Subaru telescope, the LBT, and the Keck telescopes are examples of Earth-bounded telescopes that detect infrared radiation. The HST also has capabilities to observe in the near infrared.

Because all objects, regardless of their temperature, emit infrared radiation, infrared astronomy involves the study of just about everything in the universe.

Since infrared telescopes detect heat, they have to be very cold. Otherwise, the telescopes will detect their own radiation.

Infrared radiation observes objects that are not too hot, like dying stars (red-giant) and stars in formation (protostars).

The interstellar dust is transparent to infrared wavelengths, so infrared telescopes detect the radiation emitted by objects that are inside dusty clouds. These objects are invisible to observation in the visible part of the spectrum.

All infrared images are shown in false color. Each color represents different temperatures.

Space Astronomy

The advantage of placing telescopes in orbit around the Earth or the Sun is avoiding Earth's atmosphere. The Earth is opaque to the UV, far infrared, X-ray, and gamma radiation. Therefore, these telescopes are placed in space.

Far Infrared and UV Telescopes

Water vapor and CO_2 absorb all of the far infrared radiation. The ozone layer absorbs most of the UV radiation. Therefore, these telescopes must be located in space. In August 2003, National Aeronautics and Space Administration (NASA) launched the 0.65-m infrared Spitzer Space Telescope. The telescope follows Earth in its orbit around the Sun but several million kilometer behind the Earth to avoid the extreme heat.

Several UV telescopes have been placed in orbit around the Earth. The Galaxy Evolution Explorer, Galex, is currently exploring the universe in the UV wavelength. It was put in orbit in April 2003 and has since been observing galaxies across billion of years of cosmic evolution.

Infrared astronomy studies objects that are not very hot, like old stars, protostars, and planets. UV astronomy, on the other hand, studies high temperature events.

X-Ray and Gamma Ray Observations

The Chandra X-Ray observatory, which was launched on July 23, 1999, observes the sky in the X-ray region of the electromagnetic spectrum.

X-rays are produced by very violent events and by extremely hot objects.

Other shorter wavelength radiation studied from space are gamma ray and comic rays.

The following Web site contains a gallery of images in each of the different regions of the electromagnetic spectrum: http://imagers.gsfc.nasa.gov/ems/radio.html.

Hubble Space Telescope

The HST is the jewel of astronomy. It was placed into orbit around Earth in 1990, at a distance of 600 km from the Earth. Hubble has capabilities to observe in the **visible**, **near-UV**, and **near infrared** wavelengths. It covers wavelength ranges from 100 to 2,200 nm.

The telescope is 13.1 m (approximately 39 feet) long and 4.3 m (12.9 feet) in diameter at its widest point. Its primary mirror has a diameter of 2.4 m (approximately 96 inches), which gives an angular resolution of about 0.05 arc seconds.

Power to the two on-board computers and the scientific instruments is provided by two 2.4 × 12.1 m solar panels.

The power generated by the solar arrays is also used to charge six nickel-hydrogen batteries. This provides power to the spacecraft during the roughly 25 min per orbit in which HST is within the Earth's shadow. Hubble completes one full orbit every 97 min.

Hubble has an orbital speed of 5 miles per second. If a car could travel that fast, a road trip from Los Angeles to New York City would take only 10 min.

The HST is the largest orbiting space-based telescope.

It has two cameras: one for wide field view and the other for detailed large-scale images.

> **Q27. Which of the following wavelength region does not require a space telescope to observe astronomical phenomena?**
> a. gamma (γ) ray
> b. X-ray
> c. optical
> d. far infrared
> e. they all require space telescopes
>
> **Q28. To search for a cool cloud of atomic hydrogen gas, you will use _____ telescope.**
> a. an optical
> b. an UV
> c. a gamma ray
> d. a RT
>
> **Q29. To study astronomical events that involve very high temperatures you can use**
> a. an optical telescope
> b. an UV
> c. a RT
> d. an infrared telescope

(a) (b)

FIGURE 4-8. *Two different views of The Hubble Space Telescope.*

Credit: NASA Hubble Space Telescope Collection

Q30. What might we detect with an X-ray telescope that we could not detect with an infrared telescope?

a. Radiation emitted by cool stars

b. Violent high-energy events

c. Clouds of cold gas in space

d. Both a and b above

e. All of the above

Q31. Optical telescopes are commonly used at night, but radio telescopes can be used day or night.

a. True

b. False

Q32. Your eyes are able to separate objects on the Moon that are separated as close as 0.6 arc minute as seen from Earth.

a. True

b. False

Q33. This telescope is orbiting space and makes observations in UV wavelengths to measure the history of star formation.

a. HST

b. Galaxy Evolution Explorer

c. Chandra telescope

d. Spitzer Space Telescope

The last and final mission to service the Hubble Space Telescope took place in May 2009. Over the course of five spacewalks, astronauts installed two new instruments, repaired two others, and replaced a number of components essential to the telescope's smooth functioning. It is expected that the mission has extended the operational lifespan of the telescope until at least 2014.

NASA is planning the next generation of space telescopes. The James Webb Space Telescope (JWST) will be a large optical infrared telescope with a 6.5-m primary mirror. Launch is planned for 2013.

JWST will be in orbit around Earth at a distance of 1.5 million km, well beyond the Moon's orbit. http://www.jwst.nasa.gov/ index.html.

ANSWERS

1. d
2. d
3. b
4. a
5. a
7. c
8. c
9. a, Notice that if the problem had been phrased as follows "The LGP of telescope B is times the LGP of telescope A," The correct answer would have been four.
10. d
11. a (40/20 = 2)
12. a
13. a, The larger the diameter, the better the resolving power of the telescope : 12/4 = 3.
14. a, The more the LGP, the fainter the object telescope can detect $(400/40)^2$.
15. a
16. b
17. Partially eliminates the disturbing effects of the Earth's atmosphere.
18. a
19. d
20. e
21. e
22. b
23. b
24. a
25. a
26. b
27. c
28. d, Atomic hydrogen emits 21-cm radio-radiation when the electron undergoes spin flip.
29. b, UV radiation is emitted by very hot objects.
30. b, X-ray is emitted by extremely hot objects, with temperatures of a few million Kelvin.
31. a
32. a, It is true because the human eyes are able to separate objects as close as 0.5 arc minute.
33. b

UNIT 2

The sun is the closest star to us. Like the stars, the sun formed from a collapsed molecular cloud known as the solar nebula.

By discussing the mechanism that gave rise to the solar system we discovered the process that gave origin to the stars.

This unit will describe the main characteristics of the solar system and the most likely process by which the sun and the planets were formed.

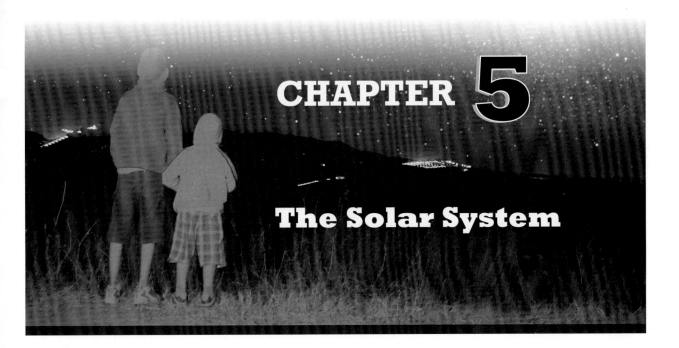

CHAPTER 5

The Solar System

INTRODUCTION

The solar system includes the Sun, planets and their moons, the asteroids, comets, and many large planet-like objects that orbit the Sun beyond Neptune in the Kuiper Belt.

In the first part of this chapter, we will study the main characteristics of the bodies that form our solar system. In the second part of this chapter, we will discuss the process by which the solar system was formed.

The Sun is the largest object in the solar system. If we represent the Sun by a small peach, then Mercury, Venus, Earth, and Mars would have the size of a grain of sugar, and Jupiter would be the size of an apple seed. Saturn, Uranus, and Neptune would be about half the size of an apple seed, and Pluto would be speck of salt. Figure 5-1 gives the approximate size of the planets and the Sun.

The planets are divided into three groups:

1. Terrestrial planets: Mercury, Venus, Earth, and Mars.

2. Jovian planets: Jupiter, Saturn, Uranus, and Neptune.

3. Dwarf planets: a new category of planets that includes Pluto, Ceres, and Iris.

Even though the Sun and the planets are quite large, the solar system is practically empty. There is almost nothing in between the orbits of the planets and the Sun—the only exception is the small belt of asteroids between the orbits of Mars and Jupiter.

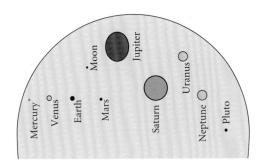

FIGURE 5-1. *Approximate relative size of planets and Sun.*

THE MAIN CHARACTERISTICS OF THE PLANETS OF THE SOLAR SYSTEM

All the objects that form the solar system orbit the Sun, except the moons of the planets which move around their respective planets. The planets move in their orbits around the Sun counterclockwise, as seen from the North Pole of Earth. Most of the orbits are almost in the same plane, with two exceptions: Mercury's and Pluto's orbits are tipped 7 and 17 degrees off the common plane, as shown in Figure 5-2. As the planets move in their orbits, they also rotate around their axes. The inclination (tilt) of their axes is generally small with respect to their plane of rotation. For example, the Earth's axis is inclined 23.5 degrees, and Mars' axis is inclined about 25 degrees. There are two exceptions to this general trend: 1) Venus rotates backwards (about 180 degrees) and 2) Uranus rotates on its side (about 90 degrees). The Sun's axis is tipped 7 degrees with respect to the plane of the ecliptic.

The period **P** of revolution and the average separation **a** of a planet from the Sun are related by Kepler's third law $a^3 = P^2$. (See chapter two).

Starting from the Sun, the planets are Mercury, Earth, Mars, the dwarf planet Ceres, Jupiter, Saturn, Uranus, Neptune, the dwarf planet Pluto, and finally, the dwarf planet Iris.

GENERAL CHARACTERISTICS OF THE TERRESTRIAL PLANETS

1. All the terrestrial planets (Mercury, Venus, Earth, and Mars) revolve close to the Sun in orbits that are closely spaced from each other, as shown in Figure 5-2.

2. The terrestrial planets are smaller and have less mass than the Jovian planets. According to size, from small to large, they are Mercury, Mars, Venus, and Earth. The radius of

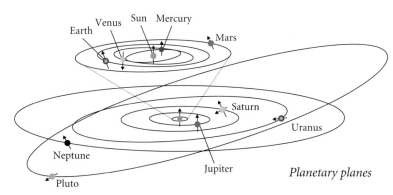

FIGURE 5-2. *Most of the planets orbit the Sun in the same plane, except the orbits of Mercury and Pluto that are 7 and 17 degrees, respectively, off the common plane.*

the Earth is 6,380 km (12,720 km of diameter). Venus is similar to Earth in mass and volume.

3. The terrestrial planets have thin atmospheres. However, Mercury does not have any atmosphere because its surface is too hot and it has a low-speed escape velocity. (Escape speed is explained shortly).

4. The average surface temperature of the terrestrial planets is relatively high. The surface temperature of Mercury during the day is approximately 700 K (800°F) and about 100 K (−280°F) during the night.

   Venus has a large greenhouse effect, which causes it to have the highest surface temperature (745 K or 882°F) of all the planets in the solar system.

   The surface temperature of Earth is on average 80°F, just right for physical comfort. Mars has surface temperatures that range between −220 and 63°F.

5. The terrestrial planets are solid and predominantly rocky. They contain high-melting materials such as Fe, Al, Ni, Si, and silicates. They also have a relatively high density, between 3.9 and 5.4 g/cm³.

   [What is density?

   It is the ratio between the mass and the volume of an object:

   density (ρ) = mass (M)/Volume (V)

   or

   $\rho = M/V$

   The density is usually expressed in kilogram per cubic meter or in gram per cubic centimeter.]

6. All terrestrial planets have metallic cores and rocky mantles. The mantle is the region between the crust (surface) and the core, or interior, of a planet. The core temperature of Earth is about 6,000 K. The core temperature of the other terrestrial planets is lower.

7. Craters are common in all terrestrial planets. The surface of the Moon and Mercury is dotted with craters. Mercury, shown in Figure 5-3, is the planet that has the greatest number of craters, which indicates that its surface is very old. Mars and Venus have less craters than Mercury but more than Earth.

   The majority of the Earth's impact craters have disappeared due to erosion, volcanic action, and the plate tectonic activity.

   If the surface of a planet or Moon is dotted with abundant craters, it tells us that that surface has been inactive for a long time. That is the case with the Moon, Mercury, and Venus. Why?

Q1. Most of the surface of this planet is covered with impact craters.
   a. Earth   b. Mars
   c. The Moon   d. Mercury
   e. b and c

Q2. Of the following planets, _____ has the highest interior temperature.
   a. Earth   b. Mercury
   c. Venus   d. Mars

Q3. Which of the following planets has the highest surface temperature?
   a. Earth   b. Mercury
   c. Venus   d. Mars

> **Q4.** Which of the following planets has the thickest atmosphere?
>
> a. Venus b. Mars
>
> c. Mercury
>
> **Q5.** Which of the following planets rotates the slowest?
>
> a. Mercury b. Mars
>
> c. Venus
>
> **Q6.** Which of the following planets rotates the fastest?
>
> a. Earth b. Mars
>
> c. Jupiter d. Venus
>
> **Q7.** Which of the following planets has liquid hydrogen in its interior?
>
> a. Earth b. Pluto
>
> c. Venus d. Jupiter

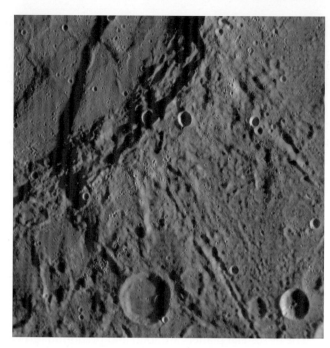

FIGURE 5-3. *The surface of Mercury has abundant impact craters. The large crater in the lower-left is about 143 miles in diameter.*

Credit: NASA Jet Propulsion Laboratory (NASA-JPL)

8. Except for Venus, which rotates backwards, the terrestrial planets rotate on their axes in the same direction.

   The rotation period of the terrestrial planets is large in comparison with the rotation period of the Jovian planets. Earth and Mars rotate in 24 h, Mercury in 84 Earth-days, and Venus in 244 Earth-days. By comparison, Jupiter, which is the largest planet in the solar system, rotates in only 10 hours.

9. The terrestrial planets have few, if any, moons. Mercury and Venus are moonless, the Earth has one Moon, and Mars has two small moons (asteroids) circling the planet in its equatorial plane.

GENERAL CHARACTERISTICS OF THE JOVIAN OR GIANT PLANETS

1. The orbits of the Jovian planets (Jupiter, Saturn, Uranus, and Neptune) are very wide and lie almost in the plane of the solar system, which is the plane of the ecliptic. The average distances from the Sun are Jupiter: 5 AU, Saturn: 10 AU, Uranus: 19 AU, and Neptune: 30 AU.

2. Their radii are several times the Earth's radius. The radius of Jupiter is 11 times the Earth's radius, the radius of Saturn is 9 times the Earth's radius, Uranus' radius is 4 times the Earth's radius, and Neptune's radius is 4 times the Earth's

radius. Jupiter, the largest and most massive planet, has a mass approximately 318 times the mass of the Earth.

3. These giant planets rotate very fast: Jupiter and Saturn have a rotation period of 10 h, Uranus a rotation period of 17 h, and Neptune of 16 h. (Compare these values with the rotation periods of the terrestrial planets.)

4. All the Jovian planets have low average density, between 0.7 and 1.7 g/cm^3. The low densities suggest that these planets are mostly rich in ammonia (NH_3), methane (CH_4), and water.

5. The Jovian planets have large hot cores. Jupiter's core is the size of the Earth and has a temperature of about 25,000 K, which is about four times the surface temperature of the Sun. The cores contain molten silicates and heavy metals, such as iron and nickel. (Rocky core with heavy metals.)

6. The Jovian planets have large atmospheres, consisting mainly of hydrogen and helium. Their cloud tops have very low temperatures. The cloud top temperature of Jupiter is 160 K, Saturn is 90 K, Uranus is 60 K, and Neptune is only 60 K. The atmospheres of some of the Jovian planets are marked with great storms, such as the Great Red Spot on Jupiter and the Great Dark Spot on Neptune. See Figure 5-4. The Jovian planets do not have solid surfaces; therefore, they do not have craters as the terrestrial planets do. Even though they have rocky cores as big as the Earth, they are called **gaseous giants** because of their huge atmospheres. On average, they are composed of about 80% hydrogen, 19% helium, and 1% other elements.

7. All the Jovian planets have ring systems. The rings of Saturn are the only rings visible from Earth and are made of ice and rocky particles. Some of these particles are small, and others are as large as a big house. Jupiter and Saturn are the largest and most massive Jovian planets.

> **Q8. Which of the following planets has the highest core temperature?**
> a. Earth b. Mercury
> c. Venus d. Jupiter
>
> **Q9. The diameter of Jupiter is about _____ times the Earth's diameter.**
> a. 5 b. 6
> c. 11 d. 18
>
> **Q10. Which of the following planets has the largest atmosphere?**
> a. Earth b. Mars
> c. Jupiter d. Venus

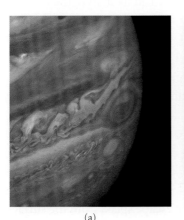
(a)

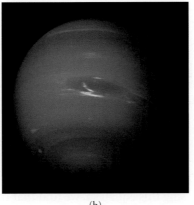
(b)

FIGURE 5-4. (a) *Jupiter's Great Red Spot.* (b) *Neptune's Great Dark Spot.*

Credit: NASA Jet Propulsion Laboratory (NASA-JPL)

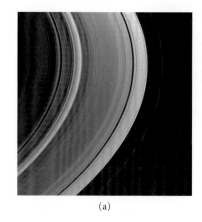 

(a)   (b)

FIGURE 5-5. (a) *The ring of Saturn.* (b) *The Moon, Mimas, of Saturn as viewed by the Cassini spacecraft 01-07.*

Credit: NASA/JPL/Space Science Institute

    Pressure in their interiors is so high that deep below their atmospheres, the hydrogen becomes liquid. The transition from gas to liquid is gradual, meaning that there is not a clear boundary between the gaseous and the liquid region.

8. All the Jovian planets have several moons. Jupiter has more than 22 moons. The four largest moons are the Galilean moons (discovered by Galileo), which are Io, Europa, Ganymede, and Callisto. Saturn has about 30 moons, many of which are small. The largest is Titan. Titan is the second-largest moon in the entire solar system. Ganymede is slightly larger.

    The space craft Cassini is studying Saturn, its rings, and its moons. Figure 5-5 shows a close-up view of the rings of Saturn and the Moon Minas. These pictures were taken by the cameras on board of Cassini.

ESCAPE SPEED

Why do some planets have atmospheres and other planets do not? To give an answer to this question, we need to analyze the concept of escape speed. The escape speed is the minimum initial speed that an atom, a molecule, and/or an object needs in order to escape from the surface of a star, a planet, or Moon.

The value of the escape speed from the surface of a planet, Moon, or star depends on its mass and radius. It is given by the following formula:

$$V_{escape} \propto \sqrt{\frac{mass}{radius}}$$

Using this equation, we find that the escape speed from the surface of the Earth is about 11.0 km/s, while the escape speed from the surface of the Moon is only 2.4 km/s. See Figure 5-6.

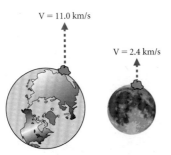

FIGURE 5-6. *Comparing the escape speeds from the surface of the Earth and the Moon.*

Credit: Moon image courtesy of JupiterImages, Inc.

The Escape Speed and the Atmosphere of the Planets

If a planet has low escape speed and high surface temperature, it likely does not have an atmosphere. Why? If the surface temperature of a planet is high, the atoms and molecules of the gases of the atmosphere might have speeds comparable to the escape speed from that planet; therefore, the gases will escape. That is exactly the case on Mercury.

What is the minimum speed that the atoms and molecules of a gas must have to escape from the surface of a planet?

Calculations show that if the speed of the atoms and molecules of a gas on the surface of a planet is equal to one-sixth the escape speed from that planet, the gas will escape. Light molecules move faster than massive molecules; therefore, the light molecules (like hydrogen) escape easier than the heavier ones (like nitrogen or carbon monoxide). The hydrogen molecules in the Earth's atmosphere have an average speed of 2.0 km/s. The more massive oxygen molecules have an average speed of 0.5 km/s. Comparing these values with the escape speed of Earth divided by 6 (11.0 km/s/6) = 1.8 km/s, we see that this value is larger than the speed of motion of hydrogen. As a result, hydrogen has escaped or is in the upper atmosphere, while the more massive and slower oxygen molecules remain in the atmosphere. This is because they move with lower speeds than 1.8 km/s. The same applies to nitrogen, carbon monoxide, and other heavier molecules.

The surface temperature of a planet also determines whether or not a planet has an atmosphere. Consider Mercury and Titan, the Moon of Saturn, both of these objects have similar sizes and masses and thus, about the same escape speed of 4.4 km/s. However, the surface temperature of Mercury during the day is approximately 700 K, while the maximum surface temperature of Titan is only 94 K. Because of the high surface temperature on Mercury, the average speed of the atoms and molecules of the gases is larger than one-sixth the escape speed from the planet. Therefore, all the gases have escaped from its atmosphere. Titan, on the other hand, has low surface temperature, and all the

> **Q11.** The escape speed from the surface of planet "X" is 24 km/s. The average speed of oxygen on this planet is 3 km/s. This information tell us that the atmosphere of planet "X" _____.
>
> a. contains oxygen
>
> b. does not contain oxygen
>
> Explain your answer
>
> **Q12.** What is escape speed?
>
> **Q13.** Our Moon does not have any atmosphere because _____.
>
> a. it has a low escape speed
>
> b. it is too close to the Sun
>
> c. of Earth gravity
>
> d. it is too far from the Earth
>
> **Q14.** Which terrestrial planet has the lowest escape speed, and how does it affect its atmosphere?

molecules of the gases have speeds lower than one-sixth the escape speed from the surface. As a result, Titan has an atmosphere. Figure 5-7 shows the surface of Titan. The surface of Titan is hidden by the clouds present in its atmosphere. The surface of Mercury can be imaged because the planet has no atmosphere. Our Moon is similar to Mercury and has no atmosphere.

FIGURE 5-7. *Surface of Titan.*

Credit: NASA Jet Propulsion Laboratory (NASA-JPL)

Mars has an average surface temperature of 38°F and an escape speed of only 5 km/s. Even though its average temperature is low, its escape speed is not enough to retain light elements such as hydrogen. Only the heavier molecules like carbon dioxide remain in the atmosphere. The Jovian planets are massive and cold. Therefore, they have huge atmospheres. Why? Practical application. To launch the Apollo spacecraft that took the men to the Moon, powerful rockets had to be used. However, the rockets used to lift the astronauts from the Moon's surface, where the escape speed is only 2.4 km/s, were much smaller and less powerful. See Figure 5-8.

(a) (b)

FIGURE 5-8. (a) *The massive Saturn-V lifts off July 16, 1969, powering Apollo 11 into orbit.* (b) *Apollo 11 lifting from the Moon.*

Credit: (a) NASA Kennedy Space Center (NASA-KSC); (b) NASA Langley Research Center (NASA-LaRC)

THE DWARF PLANETS

The dwarf planets are Pluto, Haumea, Makemake, Eris, and Ceres. The first four dwarf planets are beyond the orbit of Neptune in a large zone of ice bodies, called the **Kuiper Belt**. Ceres is inside the asteroid belt.

Pluto orbits the Sun in 248 years at a distance of 40 AU. Haumea circles the sky around the Sun in 285 years at a distance of 43 AU. Makemake does the same in 310 years at a distance of 48 AU. Eris orbits the Sun in 557 years at a distance of 67 AU.

Astronomers believe that there are other objects orbiting the Sun in the Kuiper Belt which are yet to be discovered. These new discoveries might be included in this new category of planets.

The dwarf planet Ceres, with a diameter of about 1,000 km, orbits the Sun in 4.6 years in the asteroid belt at a distance of 2.77 AU.

MAGNETIC FIELDS OF PLANETS AND STARS

The Jovian planets, the Sun, the stars, and the Earth have magnetic fields in their interiors. By studying the Earth's magnetic field, we can get a glimpse of the magnetic fields of other celestial bodies. Therefore, we will dedicate a few paragraphs to describe the Earth's magnetic field.

What generates this magnetic field of Earth?

It is believed that the magnetic field on Earth is a consequence of the **dynamo effect**. The dynamo effect is a combination of two factors: the rotation of the Earth's core and the convection of a conducting medium in the Earth's interior.

The matter in the interior of the Earth has a temperature of about 6,000 K. At this temperature, everything is liquid including the metals. Further, most of the atoms are ionized. In contrast, the middle and the top layers of the Earth are cooler. Therefore, a large amount of ionized hot matter moves upward where, after cooling, it sinks back down again to the interior. This establishes enormous convection currents of ionized matter inside the Earth. The convection motion of this ionized matter produces electric currents that move upward and downward inside the Earth. The combination of these convection currents and the Earth's rotation produce the magnetic field inside the Earth. This mechanism is called the **dynamo effect**. See Figure 5-9a.

The Earth's magnetic field extends several Earth's radii outward, which forms the **magnetosphere**, as shown in the last drawing in Figure 5-9b. The Earth's magnetosphere, which is the magnetic field around Earth, traps most of the charged particles

Q15. Mercury is similar in mass to Titan (a Moon of Saturn), and therefore, they have about the same escape speed. However, Mercury does not have any atmosphere whereas Titan does. This is because _____.

a. Escape velocity of Titan is twice that of Mercury

b. Titan is very close to Saturn, and Saturn's gravitational pull keeps the gases of the atmosphere of Titan in place

c. Titan's average surface temperature is about 100 K, while Mercury's surface temperature can reach 880 K degrees during the day

d. Titan has liquid water running on its surface while Mercury does not

e. a and b

coming from the Sun in the solar wind. (The solar wind consists of gentle streams of charged particles (or plasma) consisting mainly of protons and electrons. It comes from the Sun's atmosphere.)

The charged particles from the Sun are trapped in the magnetosphere in the so-called **Van Allen belt** region. The trapped particles spiral along the magnetic field lines toward the Earth's poles. The interaction of these charged particles with Earth's atmosphere produces the aurorae, or the northern and southern lights, which are visible in the northern and southern latitudes.

The dynamo effect in the Jovian planets is stronger than that in Earth for two reasons:

1. These planets have a hotter and larger core. Therefore, they have larger convection currents of ionized matter in their interiors and

2. they rotate much faster.

If a magnetic field is detected on the surface of a planet, we conclude that the core of that planet is molten and hot. Why?

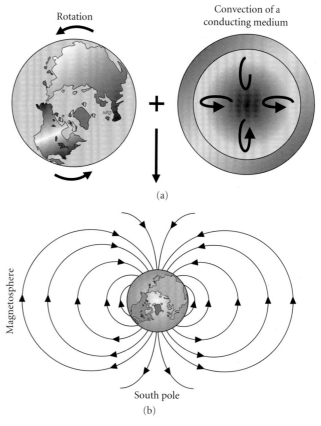

FIGURE 5-9. *The Earth's magnetic force is produced by the dynamo effect.*

Chapter 5—*The Solar System*

Space Debris

Space debris includes everything in the solar system that is not the Sun, planets, or moons. Space debris includes asteroids, meteoroids, meteorites, meteors, and comets.

Asteroids

Asteroids are rocky solid objects which orbit the Sun in the asteroid belt (see Figure 5-10) between the orbits of Mars and Jupiter at a distance approximately between 1.8 and 3.0 AU. According to one leading hypothesis, the asteroids of the asteroid belt are the debris left by a planet that failed to form due to the disturbing gravitational effects of Jupiter.

The orbits of most of these asteroids have small eccentricity, but a few have elliptical orbits that take them into the inner solar system, where they could potentially collide with Mars or Earth. These are the **Apollo asteroids** or **Earth-crossing asteroids**. Others, such as the Trojan asteroids, share the orbit with Jupiter. See Figure 5-10.

Asteroids come in different sizes—from small pebbles to large objects of several hundred kilometers in diameter. There are 26 known asteroids larger than 200 km in diameter, approximately 200 asteroids with a diameter of 100 km (60 miles), tens of thousands of asteroids with diameters larger than 10 km (roughly 6 miles), and probably billions with diameters smaller than 1 km (0.6 miles).

The dwarf planet, Ceres, has a diameter of about 1,000 km (625 miles). This object is the largest object in the asteroid belt and contains 25% of the mass of all the asteroids combined.

> **Q16. According to one theory, the asteroids in the asteroid belt are the remains of a planet that failed to form at a distance of 2.8 AU from the Sun about 4.5 billion years ago.**
>
> a. True   b. False
>
> **Q17. All known asteroids circle the Sun in the asteroid belt between the orbit of Mars and Jupiter.**
>
> a. True   b. False

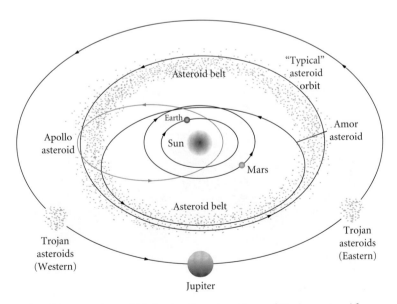

FIGURE 5-10. *Asteroid belt, Apollo asteroids, and Trojan asteroids.*

> **Q18. Which of the following is *not* a characteristic of any of the asteroids?**
>
> a. Irregular shape
> b. Solid
> c. Rotating on their axes
> d. Small atmospheres

All asteroids have irregular shapes, are heavily cratered (see Figure 5-11), and rotate on their axes, as shown in the Hubble Space Telescope composite image (Figure 5-12).

Our telescopes do not have enough resolution to observe surface details on the asteroids. Even the surface of Ceres, when viewed through the lens of the Hubble Space Telescope, appears smooth, without craters or irregularities. The spacecraft named Galileo, on its way to Jupiter, provided the first close-up view of the surface of the asteroids Mathilde, Gaspra, and Ida. See Figure 5-11.

Although the asteroids are rocky in composition, their density is only approximately 1.3 g/cm³. Their low average density implies that they are not made of one single piece of compacted hard solid rock but are thought to be large chunks of rocks held together by their mutual gravitational force.

FIGURE 5-11. *Three asteroids, as seen by the Galileo spacecraft on its way to Jupiter.*

Credit: NASA Jet Propulsion Laboratory (NASA-JPL)

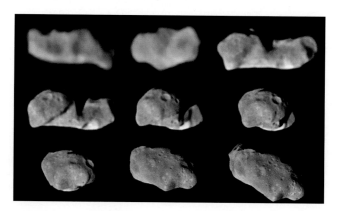

FIGURE 5-12. *The Hubble Space Telescope shows that Ida rotates on its axis.*

Credit: NASA Jet Propulsion Laboratory (NASA-JPL)

In general, a body of rocky composition is a solid body that contains silicates, SiO_2, and heavy metals such as Fe, Ni, Al, and Mg.

A few asteroids have moons, like the asteroid Ida. Even more surprising, observations indicate that the asteroid Vesta, 525 km across, has traces of lava flow on its surface. This indicates that at some point long ago, it might have had volcanic eruptions.

On February 12, 2001, the NEAR spacecraft landed on the surface of Eros, a 34 km rocky asteroid.

On September 2007, the spacecraft Dawn was launched to explore the two most massive and largest bodies of the asteroid belt: the asteroid Vesta and the dwarf planet Ceres.

Comets

When comets approach the inner solar system, the solar radiation sublimes (vaporizes) material from their nuclei, forming long tails that reflect the light from the Sun and make them visible.

Most of the comets reside far away from the Sun in the Oort cloud and in the Kuiper Belt. The Oort cloud is a spherical shell surrounding the solar system at a distance between 10,000 and 50,000 AU (approximately 0.5 light years). See Figure 5-13. This region is the seat of trillions of small frozen (solid) bodies of about 10–20 km in diameter. These cold frozen bodies are the **nuclei of comets**. These frozen bodies are made of a mixture of ices and silicates. The ices contain frozen water and frozen gases, such as ammonia (NH_3), methane (CH_4), carbon dioxide (CO_2), as well as a mixture of tiny dusty solid particles of silicates. Gravitational disturbances of occasional passing stars may perturb the orbits of some of the frozen comets in the Oort cloud. This causes the comets to begin a journey toward the inner solar system.

The comets from the Oort cloud have a period of reappearance in the inner solar system of more than 200 years. Because of this, they are known as the **long-period comets**. The astronomer, John Oort, while analyzing the orbits of long-period comets in 1950, noticed that these comets approached the solar system in different directions, with random elliptical orbits that were not in the planetary disk. From his observations

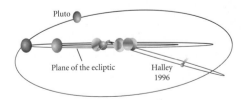

FIGURE 5-13. *The orbit of the comet Halley is below the plane of the ecliptic.*

and calculations, he concluded that these comets originated in a spherical shell surrounding the solar system. This spherical shell extends from 10,000 to 50,000 AU and became known as the **Oort cloud**.

The **short-period comets**, with periods of less than 200 years, come from the Kuiper Belt. The Kuiper Belt is a region of icy planet-like objects lying beyond the orbit of Neptune in the same plane of the solar system.

The Kuiper Belt contains billions of frozen potential comets and other larger bodies. The composition of these frozen comets is similar to those in the Oort cloud. The comets in this region normally move in orbits around the Sun, but the gravitational influence of Neptune, or the collision of two comets, might disturb the normal orbit of a few of them. This causes the frozen comets to follow elliptical orbits that bring them close to the Sun.

The Halley comet, with a period of 76 years, is the most well-known comet coming from the Kuiper Belt. See Figure 5-13. When a comet approaches the inner solar system, the heat from the Sun vaporizes matter from the nuclei. This forms a coma and a hydrogen envelope and develops two tails in well-developed comets.

The tails always point away from the Sun and can be as long as 150 million km.

The **ion** or **gas tail** is formed by ionized gases and is driven away from the nucleus by the solar wind. Most of the ions are ionized atoms of carbon dioxide, carbon monoxide, hydrogen, oxygen, carbon, and also ionized water molecules. The tail looks blue because the carbon monoxide ions scatter the Sun's blue light more efficiently than the other colors.

The dust tail is carried away from the nucleus by the Sun's radiation pressure. The tail looks white because the dust particles reflect the white light (all colors) from the Sun. The coma surrounds the nucleus and contains dustand gases that have evaporated from the nucleus. It can have a diameter of about 100,000 km. The coma has an invisible envelope of hydrogen.

Only a few of the comets that come to the inner solar system are visible to the naked eye.

The Nuclei of Comets

The Giotto spacecraft gave us a direct glimpse at the heart of the Halley comet in 1986. The Stardust spacecraft took high-resolution photographs of the comet Wild in 2004. From these observations, it is believed that the nuclei of comets are a mixture of pulverized dusty rock (silicates) ammonia, methane, carbon dioxide, and other ices, all loosely packed in a small volume between 10 and 20 km across a density of only 0.3 g/cm^3 (one-third the density of water). The temperature of the nuclei is very low, but the surface of the side closer to the Sun gets warmer and vaporizes, giving birth to the coma and tails.

When the comets come close to the inner solar system, they leave behind a trail of small particles that, later on, produce the meteors and the meteor showers.

The Greeks thought that nothing changed beyond the Earth's atmosphere. Therefore, they believed that the comets were part of the Earth's upper atmosphere. Tycho Brahe, in 1577, was the first person to explain that the comets were produced far away from the Earth as part of the solar system. Current theories suggest that both comets and asteroids are left over from the formation of the solar system.

Meteors and Meteorites

Within the solar system, there are billions of small solid particles (tiny bits of metal and rock) moving in different directions. These particles are called **meteoroids**. Some of these particles come from the asteroids and/or the comets. When these particles enter the Earth's atmosphere, the friction with the molecules of Earth causes them to reach very high temperatures.

The particles coming from the asteroids are hard and compact. Therefore, they can survive high temperatures and make it to the surface of the Earth. These are **meteorites**, and thousands of them have been found in various places.

The particles coming from comet debris are fluffy and porous and do not endure the high temperatures caused by friction. Therefore, they burst into incandescent vapor, producing bright trails of light or meteors. These are commonly called "**shooting stars**."

When the Earth's atmosphere goes through the orbit of a Comet, it encounters several tiny particles left behind by the comet, giving rise to meteor showers. See Figure 5-14.

The faster the meteors move, the brighter they are. After midnight, the meteors move in the same direction in which the Earth rotates. So, the best chance to see a meteor is after midnight. On

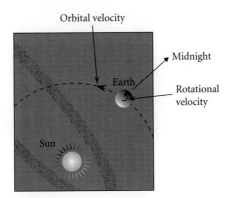

FIGURE 5-14. *Origin of a meteor shower.*

Chapter 5—*The Solar System*

Q19. The comets in the Oort cloud _____.

  a. have small tails
  b. are solid balls of frozen ices
  c. are visible from Earth with powerful telescopes
  d. have small atmospheres

Q20. When we say that the nuclei of the comets contain "ices," we mean that they contain frozen _____.

  a. water
  b. methane
  c. ammonia
  d. carbon dioxide
  e. all the above

Q21. Most potential comets reside in the _____.

  a. asteroid belt
  b. Kuiper Belt
  c. Ort cloud
  d. interstellar space
  e. b and c

Q22. The tails of comets point toward the Sun when they are approaching the Sun and away when they are receding from the Sun.

  a. True   b. False

Q23. The nuclei of comets mainly consist of _____.

  a. a solid ball of ice
  b. a mixture rocks in the form of dust and ices
  c. low-temperature hydrogen gas
  d. low-temperature carbon dioxide gas

any moonless night, we can see a few meteors per hour. Each time a comet revisits the solar system, it loses matter which forms a trail of tiny particles.

A comet may last only 100–1,000 orbits around the Sun. All that may be left of the comet will be a pile of rubble.

Every year several comets approach the Sun, but most of them are invisible. Ocasionally the comets collide with planets, moons, and the Sun.

In summary, most of the meteors, or "shooting stars," are pieces of comets. Meteorites are mainly pieces of asteroids or even planets. Only a few sporadic meteors are not associated with comet debris and comet orbits.

Planetary Impacts

Small impacts of meteorites with Earth are common. Large impacts are possible and have occurred in the past. On Earth, most traces of larger impacts have been erased by erosion. The surfaces of Mercury, Mars, and the Moon have abundant craters, which are silent witnesses of earlier planetary impacts. See Figure 5-15.

What would a large impact do?

Impact on land or sea makes no difference. The impact would excavate a large amount of rock. The rock will be pulverized, heated, and lofted to the upper atmosphere and into a low orbit around Earth. Hot debris would fall all around the world, triggering forest fires. The dust in the atmosphere would block sunlight for about a year. Plants and larger animals would die.

About 65 million years ago, a giant impact produced mass extinction (including dinosaurs). About 50 million years ago, a 50-m asteroid formed the Barringer Meteorite Crater in Arizona.

During the summer of 1994, the fragmented comet, Shoemaker-Levy, fell on Jupiter's atmosphere.

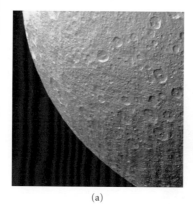

(a)   (b)

FIGURE 5-15. *The surface of Mercury and the Moon are covered with craters formed by impacts with asteroids.*

Credit: (a) NASA Jet Propulsion Laboratory (NASA-JPL); (b) NASA/JPL/Space Science Institute

Dating of Meteorites

It is generally accepted that meteorites are left over from the formation of the solar system. Therefore, by knowing their age, the age of the solar system can be obtained. The age of the meteorites is found from radioactive decay studies.

When matter solidifies, it incorporates known percentages of chemical elements. A few of them are radioactive and will decay with time. The time that it takes for half of the mass of a radioactive element to decay is called **half-life**.

If you start with a pure sample of a radioactive element, half of that sample will have decayed after one half-life has elapsed.

Table 5-1 shows the half-life of three different elements.

For example, if a rock has 20,000 particles of uranium, 10,000 of them will decay into lead within 4.5 billion years. If you wait another 4.5 billion years, 5,000 more will decay, and so on.

The age of the meteorites can be found by comparing the original amount of radioactive (unstable) atoms and stable or decayed abundance.

Figure 5-16 shows the decay of a radioactive sample that has a half-life of 1 billion years. If the initial number of radioactive particles was 700 after a half-life, or 1 billion years, 350 particles have decayed and 350 have not decayed. How many particles have decayed after three half-lives? How many have not decayed?

Radioactive decay has shown that meteorites and Moon rocks brought home by the Apollo astronauts are roughly 4.6 billion years

TABLE 5-1. *Half-life of a few elements.*

PARENT	DAUGHTER	HALF-YEARS
^{238}U	^{206}Pb	4.5 billion
^{40}K	^{40}Ca, ^{40}Ar	1.3 billion
^{87}Rb	^{87}Sr	47 billion

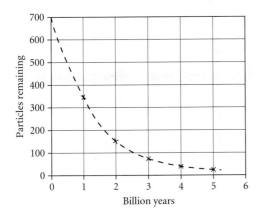

FIGURE 5-16. *The half-life of this particle is 1 billion years.*

Q24. Define the term "half-life" as applied to a radioactive element.

Q25. The decay curve of tritium is shown in Figure 5-23. What is its half-time in years?

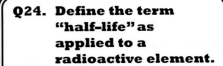

a. 98   b. 49
c. 25   d. 12

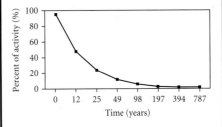

FIGURE 5-23. *Decay curve for tritium.*

Q26. The oldest rock found on the surface of Earth is about _____ years old.

a. 4.8 billion
b. 5.0 billion
c. 4.5 billion
d. 4.0 million

old. Since everything in the solar system was formed at the same time, the solar system was formed about 4.6 billion years ago.

The age of the Earth is 4.6 billion years. Nevertheless, according to radioactive dating techniques, the "ages" of the oldest rocks found on Earth are only about 4.0 billion years old. So, what is wrong?

Is the Earth younger than the other objects of the solar system?

No, it is not. The oldest rocks on the surface of the Earth have been destroyed by the motion of the tectonic plates, volcanoes, erosion, and the actions of animals and men.

Now we will focus our attention on the process by which the solar system was formed.

FORMATION OF THE SOLAR SYSTEM

In this second part, we are going to study the most probable process by which the Sun, the Earth, and the planets were formed. A similar process has given rise to the formation of the stars.

The story begins when a molecular cloud collapses under its own weight, in what is called **gravitational collapse**, to form a planetary nebula.

The Solar System and Molecular Clouds

The story of the solar system is linked to a molecular cloud, which is a large accumulation of gases and dust. [They are called **molecular clouds** because the hydrogen gas is in the form of molecules instead of atoms.] Molecular clouds are part of even larger systems called **giant molecular clouds** (GMC). Some GMC have diameters of 50 pc (approximately 150 ly). The composition of the molecular clouds is similar to the composition of the Sun and other stars. They consist mainly of (by mass) approximately 72% hydrogen and 27% helium, and the remaining 1% of heavier elements. Additionally, about 1% of the mass of the molecular clouds is in the form of dusty grains that contain heavier atoms such as silicates (rocks) and metals like iron, aluminum, nickel, and magnesium.

The dust grains in the molecular clouds are important for two reasons:

1. They are the "seeds" where atoms condense out of the cloud to form planets.

2. They absorb the radiation emitted by the stars that are inside the molecular cloud, preventing the gas from boiling away.

The molecular clouds are cradles of stars. The young stars are sources of ultraviolet and infrared radiation. Without the dust particles to absorb this radiation, the molecular clouds would evaporate after a few stars have formed. Near the stars, the temperature of the molecular cloud is high and it glows, producing the HII regions. These regions can be observed with optical telescopes. But far away from the stars, the average temperature is only approximately 10 K. Therefore, it is difficult to observe a molecular cloud in its entirety. However, the molecular clouds contain traces of carbon monoxide (CO) and other molecules that emit radio radiation which is used to map them. If astronomers detect the presence of CO in a given region of the galaxy, they assume that they have detected a molecular cloud because CO molecules are always inside molecular clouds.

Molecular clouds are the only places where stars form. The different molecules of the molecular clouds move in different directions with different speeds. The net effect of this motion causes the molecular cloud to slowly rotate around its axis. See Figure 5-17. The internal motion of the molecules also produces an outward pressure that tries to expand and destroy the molecular core. This outward force is balanced by the inward gravitational pull. Under these two forces, the molecular cloud is in hydrostatic equilibrium. See Figure 5-18.

> **Q27.** The molecular clouds have, by mass, about _____% of hydrogen, 27% of _____, 1% of _____, and 1% of dust.
>
> **Q28.** The composition of the molecular clouds is very similar to the composition of _____.
> a. Mercury
> b. the Earth
> c. the stars
> d. Ceres

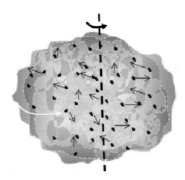

FIGURE 5-17. *Molecular clouds are very cold and slowly rotate on their axes.*

In general, the molecular clouds that are in hydrostatic equilibrium have low temperatures. However, if the temperature is too low, the gravitational force will dominate and the cloud will collapse.

Usually the molecular clouds contain a large number of dense and cold cores that give rise to the formation of star clusters. Our Sun was formed when one of theses cores collapsed under the gravitational pull.

For a molecular cloud to collapse, the hydrostatic equilibrium has to be destroyed by an external agent. What causes the gravitational collapse of a molecular cloud?

FIGURE 5-18. *Molecular clouds are in equilibrium under the action of the gravitational force (in) and the force due to pressure (out).*

The gravitational collapse is usually triggered by a shock wave.

Shock waves are traveling pressure waves. The shock waves might be produced by one of the following events:

1. A nearby supernova explosion.
2. The birth of a nearby massive star.
3. The passage of a star close to the molecular cloud.
4. When a molecular cloud enters the arms of the Milky Way. (The arms of the Milky Way are regions of slowly traveling shock waves.)

As a molecular cloud collapses, it gradually increases its temperature and its speed of rotation.

As matter falls, it converts potential energy into kinetic energy, and therefore, the atoms gain speed and temperatures increase. The higher the speed, the higher the temperature. [Recall that the average kinetic energy of the atoms in a gas determines the temperature of the gas.]

The speed of rotation increases because the core contracts. The physical law that explains this behavior is known as the **law of conservation of angular momentum**.

Astronomers call this law the skater effect because figure skaters gain rotation speed by drawing in their arms (this reduces their average radius), and lose rotation speed by extending them (increasing the radius). See Figure 5-19.

Most of the matter of the collapsing molecular cloud core accumulates in a hot center, where the Sun will form. The rest forms a flattened disk around the hot center. The entire collapsed rotating system is called a **solar nebula**. See Figure 5-20.

The rotating central bulge is called the **protosun**. It will give birth to the Sun. The disk is called the **protoplanetary disk** because planets will form there.

The temperature of the solar nebula is higher when it is closer to the protosun and lower as distance increases from the center. The

FIGURE 5-19. *Example of conservation of angular momentum.*

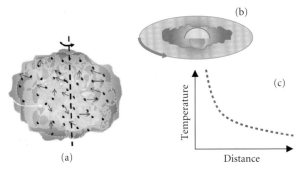

FIGURE 5-20. (a) *Uncollapsed molecular cloud.* (b) *Solar nebula and* (c) *temperature gradient of solar nebula.*

smooth distribution of temperature with distance is called **temperature gradient**. See Figure 5-20c.

THE FORMATION OF PLANETS

The temperature gradient in the solar nebula causes elements to distribute according to their melting points. Near the protosun, where the temperature is high, only refractory, or high temperature, melting elements (metals, oxides, and silicates) remain. Only few hot gases also remain in this region. This is the zone where the terrestrial planets form. The Jovian planets form in the low-temperature zone of the solar nebula which is rich in low-melting material, such as ices, gases, some hydrogen compounds, methane, and ammonia. High-melting materials are also found in this region.

The fine solid dust particles initially present in the collapsing molecular cloud core are distributed allover the entire solar nebula.

Thus, the temperature gradient explains the existence of the terrestrial and Jovian planets.

The zone that separates the two kinds of planets is usually called the **frost line**, and it is somewhere beyond the asteroid belt where the temperature considerably drops between Mars and Jupiter.

Formation of the Terrestrial Planets

The most likely steps that led to the formation of terrestrial planets were condensation, accretion, formation of protoplanets, and differentiation.

Step One: Condensation

As the solar nebula cooled down, solid particles began to condense out of the nebula around the dust particles, just as snow flakes condense in our atmosphere. The dust particles acted as condensation nuclei or seeds. Condensation is initially very effective, but it gets less efficient as the grains grow bigger. Only very tiny particles were formed by condensation.

> **Q29. Near the young "protosun" mainly _____ and other high-melting materials condensed out of the solar nebula.**
>
> a. oxygen
>
> b. low-melting material
>
> c. silicates (rocks)
>
> d. ices
>
> **Q30. Far away from the young "protosun," both low-melting materials and _____ condensed out of the solar nebula.**
>
> a. silicates
>
> b. high-melting materials
>
> c. iron
>
> d. all the above

Chapter 5—*The Solar System*

> **Q31. What is temperature gradient?**
>
> a. The change of temperature with density
> b. The change of temperature with distance
> c. The decrease of temperature with time
> d. The change of temperature with mass
>
> **Q32. As the molecular cloud core collapsed under its own gravity _____.**
>
> a. its temperature and speed of rotation increased
> b. its mass increased
> c. it cooled down and spun up
> d. it cooled down and spun down
>
> **Q33. As the solar nebula collapsed, it formed a rotating disk around the young protosun because _____.**
>
> a. the overall direction of motion of the particles in the molecular cloud was conserved as it collapsed
> b. the molecular cloud lost mass as it collapsed
> c. the molecular cloud got cold
> d. the molecular cloud got a vertigo as it collapsed

The temperature profile (temperature gradient) across the nebula determined the type of material that condensed. Close to the protosun, where the temperature was high, only high-melting materials, metals such as Fe, Ni, Al, silicates, and oxides, survived and condensed out of the nebula to give rise to the terrestrial planets: Mercury, Venus, Earth, and Mars. These planets formed inside the frost line, which is located between the orbits of Mars and Jupiter just outside the asteroid belt.

Step Two: Accretion and Formation of Planetesimal

The condensation process formed the first small clumps of matter in the solar nebula. Everything is very dynamic, so the tiny clumps kept growing by soft collisions and by sticking to other clumps. The process of growth by soft collisions and sticking together is called **accretion**, much like making a large snowball by rolling a smaller snowball in snow. Accretions build on condensation to form larger particles. The largest particles are called **planetesimals**, literally meaning pieces of planets. By accretion, billions of small and a few large planetesimals form. All the planetesimals, small and large, circled the protosun, in almost the same direction and in almost the same plane. This is the plane of the solar system or the plane of the ecliptic.

Step Three: Formation of Protoplanets

As the planetesimals circled the protosun, soft collisions contributed to the formation of larger objects. Occasionally there were head-on collisions, but these collisions shattered the planetesimals and did not aid the process of growth. The fragments of these collisions were gravitationally attracted by the larger planetesimals, helping them to grow even bigger. The larger planetesimals attracted more fragments and grew faster than the smaller ones. Eventually four large planetesimals formed inside the frost line, giving origin to the terrestrial protoplanets. The planetesimals that were not captured by the protoplanets escaped to form the asteroid belt, which populate the Kuiper Belt and the Oort cloud.

Once the protoplanets—precursors of planets—were formed, they did not grow any further because the surrounding material was already depleted.

Step Four: Density Differentiation

The protoplanets that gave origin to the terrestrial planets were as big as the planets. They are not planets yet because they were homogeneous in composition. Conversely, a full-fledged planet will have a layer structure of different composition. See Figure 5-21.

The young protoplanets were melted by the heat that was released as planetesimals fell to their surface. This heat was liberated by radioactive decay in their interior. During the period of time in which the protoplanets were melted, the elements were distributed according to their density.

The more dense materials, like iron nickel and some silicates, formed the core. On top of the core, lighter materials accumulated,

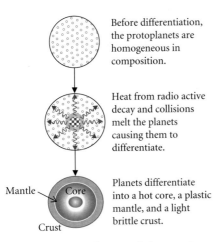

FIGURE 5-21. *Differentiation of terrestrial protoplanets.*

forming a cooler mantle (a mixture of metals and silicates) and on top of the mantle, a cool crust of light materials (rich in silicates) formed. When the protoplanets cooled down, the materials remaining locked in different layer.

At the end of the differentiation period, the planets were cooler and no longer homogeneous. The terrestrial planets were completed in about 50 million years, but they continued to accrete solid debris for another 100 million years. This was the time of intense bombardment and cratering.

Formation of the Jovian Planets and their Moons

The Jovian planets formed in the cooler region of the solar nebula, beyond the frost line. The theory to explain the formation of the Joviain planets is not firmly established.

One model proposes that the protocores that gave rise to the Jovian planets formed by a similar process by which the protocores of terrestrial planets formed. These protocores were about twice the size of the Earth and had enough gravitational force to draw material directly from the nebula to form large atmospheres.

The terrestrial protoplanets were not able to form atmospheres from the nebula because gases were not abundant where they formed.

Where the Jovian protoplanets formed, the nebula had abundant matter, and they accreted large rotating disk of gas, dust, and heavy elements where the large moons of the Jovian planets evolved.

Some of these moons are as large as Mercury and several of the moons, like Titan, have atmospheres. The largest moon in the solar system is Ganymede, a Moon of Jupiter. In size, this is followed by Titan, a Moon of Saturn.

The smaller moons of the Jovian planets were probably captured asteroids.

Q34. Which of the following is true?

a. The protoplanets were formed by condensation and small particles by accretion.

b. The planetesimals were formed by condensation and dust particles by accretion.

c. Dust particles were formed by condensation and tiny particles by accretion.

d. Dust grains were the seeds around which matter condensed out of the nebula.

Q35. The _____ in the solar nebula explains the existence of the terrestrial and Jovian planets.

a. conservation of angular momentum

b. temperature gradient

c. conservation of energy

d. mass gradient

Q36. The planets nearest the Sun contain only small amounts of _____.

   a. materials that condensed at low temperatures
   b. materials that condensed at high temperatures
   c. metals and silicates

Q37. The dust grains in the molecular clouds provide the sites for the condensation of matter out the solar nebula.

   a. True   b. False

Q38. In what part of the nebula did silicates and metals condensed?

   a. far away from the Sun where the Jovian planets formed
   b. in the Oort cloud region
   c. close to the Sun inside the frost line
   d. a and c

Q39. Why are there terrestrial and Jovian planets? Because the solar nebula _____.

   a. was cold near the Sun and hot far away
   b. was hot near the Sun and cold far away
   c. contained low- and high-melting materials that condensed at different temperatures
   d. b and c

The terrestrial planets did not form disks around them. Therefore, they did not form any moons like the Jovian planets. The two small moons of Mars were captured asteroids.

Our Moon probably formed when a planetesimal the size of Mars slammed into our planet. The collision blasted a geyser of hot gas and molten rock into orbit around Earth. The material quickly cooled and coalesced to form the Moon. The Earth must have been very different before the collision.

Another theory for the formation of the Jovian planets proposes that the outer solar nebula contained instabilities that triggered the initial collapse, which created the seeds to form the planets. This theory is known as the **gravitational collapse**. In this model, the Jovian planets did not go through condensation and accretion because they grew directly from the nebula by gravitational collapse.

The Jovian planets grew faster and bigger than the terrestrial planets because they were able to attract material directly from the nebula. They were completely formed in about 10 million years.

Most models dealing with the formation of the solar system propose that the planets formed in about the same orbits they now have. However, Neptune and Uranus formed closer to Jupiter's orbit. Gravitational interaction with Jupiter pushed them outwardly.

As the planets were forming in the disk, the center of the nebula where most of the matter had accumulated was contracting and increasing its temperature and internal pressure. This early stage is known as **protosun**. The temperature and pressure increased faster in the protosun's interior than in the protophotosphere (protosurface). When the core's temperature was 10 million K, the young protosun (protostar) began to make energy by fusing hydrogen into helium. After the ignition of hydrogen, the Sun kept contracting until its core temperature reached 15 million K. This is the Sun's core temperature as of today.

The Sun formed in about 50 million years, giving enough time for the planet formation to occur.

Atmospheres of Planets

The atmospheres of the Jovian planets were drawn directly from the solar nebula. They have not evolved and thus are primary atmospheres. The atmospheres consist mainly of hydrogen and helium, with traces of methane and ammonia.

The Jovian planets have a large escape speed, and therefore, they have large atmospheres.

The atmospheres of the terrestrial planets have different origins. Their primary atmospheres were the result of out-gassing and volcanic eruptions from the planets themselves. Primary atmospheres also were the result from gases released by collisions with planetesimals, meteorites, and comets. These atmospheres were rich in carbon dioxide and other toxic compounds.

Mercury lost its atmosphere because of its low escape speed and high average surface temperature. Venus, Mars, and Earth have a larger escape velocity than Mercury and were able to retain their atmospheres.

These primordial atmospheres were hot and rich in carbon dioxide and other toxic compounds.

The primary atmosphere of Earth evolved into a secondary atmosphere because as it cooled down, it rained, and the rain dissolved the carbon dioxide, depositing it in the bottom of the sea. This is how limestone was formed. The formation of a secondary atmosphere was further enhanced by photosynthesis and other processes.

Venus' atmosphere consists of about 95% carbon dioxide (CO_2), but it never cooled down and, as a result, never rained and eliminated its carbon dioxide. The consequence of this is that Venus developed a large greenhouse effect that has increased its surface temperature to about 780 K. Venus, after the Sun, has the highest surface temperature in the solar system. The atmospheric pressure on the surface of Venus is about 90 times larger than on Earth!

Venus is always covered with clouds. These clouds contain sulfuric acid and water droplets. The clouds also reflect the Sun's light very efficiently.

The atmosphere of Mars is rich in carbon dioxide, but due to its lower escape speed than Venus and Earth, it has been gradually loosing its atmosphere. Its atmospheric pressure is so low that water, if it ever had any, evaporated.

The surface of Mars is desert-like and has a temperature of 180 K during the night and 270 K during the day.

NASA has been extensively exploring the surface of Mars since 1997.

> **Q40. How much of the solar nebula consisted of elements heavier than hydrogen and helium?**
> a. About 90% by mass
> b. About 50% by mass
> c. About 25% by mass
> d. About 10% by mass
> e. About 1–2% by mass
>
> **Q41. If the escape velocity of Earth is similar to Venus' why are the atmospheres so different?**

The Greenhouse Effect on Earth and Global Warming

A fraction of the radiation coming from the Sun makes it through the Earth's atmosphere and reaches the Earth's surface. The atmosphere reflects about 33% of the energy that comes from the Sun. The percentage of the solar energy reflected back to space by the atmosphere of a planet is known as the **albedo**.

The surface of the Earth reradiates the absorbed energy. A portion of the reemitted energy is absorbed by the water and carbon dioxide of the atmosphere, and the rest escapes into space.

The partial trapping of the solar radiation is known as the **greenhouse effect**. Because of this effect, the temperature of the surface of the Earth does not change much from the day to the night. The atmosphere, acts as a blanket that keeps the Earth warm during the night. [Mercury does not have any atmosphere, and the temperature during the day is about 700 K and during the night only 100 K. Planets with no atmosphere have extreme atmospheres.]

On Earth, the very delicate balance produced by the greenhouse effect for millions of years is being changed by us. In the last century, we have dramatically increased the concentration of water vapor and carbon dioxide in the atmosphere. See Figure 5-22.

The average surface temperature of Earth is dangerously increasing.

Clearing the Nebula

The solar nebula that gave origin to the solar system was inside a big shroud of dust that hid the entire process from external observers. When the Sun became hot and luminous, the combined action of the solar wind and the solar radiation pressure pushed away most of the atoms of gas and dust particles that did not participate in the formation of the planes. [The solar wind is

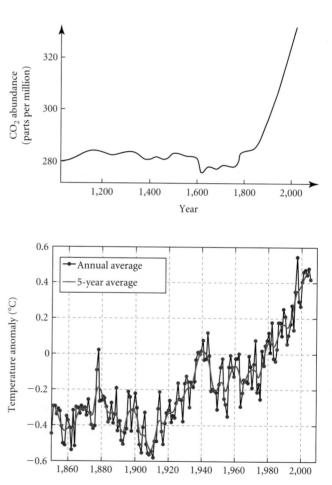

FIGURE 5-22. *The concentration of CO_2 and the average temperature on the surface of the Earth have dramatically increased in the last hundred years.*

a continuous flow of tenuous ionized gases moving outward from the Sun's upper atmosphere.]

Many of the planetesimals that were left over after the planets were formed were attracted by the young planets. This was the epoch of a heavy bombardment period that produced many of the craters that are visible on the surfaces of the terrestrial planets.

It is believed that the asteroid Belt, the Kuiper Belt, and the Oort cloud were populated by planetesimals that were gravitationally ejected by the large Jovian planets.

THE SOLAR NEBULA HYPOTHESIS AND THE PROPERTIES OF THE SOLAR SYSTEM

The solar nebula hypothesis is able to explain the key properties of the solar system:

1. The planets orbit around the Sun in the same plane and in the same direction because they were formed from the same rotating flattened disk.

2. The orbits of the planets are close to being circular because the protodisk of the solar nebula, where they formed, was approximately circular.

3. There are two main types of planets: terrestrial (rocky) and Jovian (gaseous). This is because the temperature gradient determined the type of elements along the nebula. Different elements condensed at different temperatures and distances from the Sun.

4. The common age of the solar system is easily understood because everything formed from the same solar nebula.

5. The existence of space debris is explained because the asteroids in the asteroid belt are the remains of the material that did not form a planet between Mars and Jupiter. And the ejection of planetesimal by the planets populated the Kuiper Belt with ice planetesimal and comets, and the Oort cloud with comets.

6. The presence of moons is also explained in this theory. Some moons were captured and the larger moons of the Jovian planets formed around the protoplanets.

7. The craters observed on planets and asteroids are the products of collision with planetesimals when the solar system was young.

The large tilt of rotation of Uranus, the backward rotation of Venus, and the disappearance of Mercury's crust are due to collisions with large planetesimals when they were young.

Q42. A successful scientific model of the origin of the planetary system must be able to explain which of the following features?

a. the tilt of Venus

b. the roughly circular planetary orbits

c. the roughly coplanar planetary orbits

d. the extremely distant orbit of the comets

e. all the above

Q43. How did the solar nebula get cleared of the material left over after the formation of the planets?

a. The radiation pressure of sunlight pushed gas particles outward.

b. The intense solar wind of the youthful Sun pushed gas and dust outward.

c. The planets swept up gas, dust, and small particles.

d. Close gravitational encounters with Jovian planets ejected material outward.

e. All the above.

Q44. A planet that contains hydrogen in the liquid metallic state far below its atmosphere is _____.

a. Mars
b. Venus
c. Jupiter
d. Mercury
e. it is impossible for a planet to have liquid hydrogen.

Q45. Select the false definition out of the list below:

a. dusty and icy

   nuclei of comets

b. Oort cloud

   long-period comets

c. asteroids

   not all of them lie in the asteroid belt

d. comet tails

   always point toward the Sun

e. meteor shower

   occurs when Earth crosses the debris-filled orbit of a comet

---

In conclusion, the following is a summary of the formation of the solar system (the solar nebula theory):

1. A slowly rotating cloud of gas and dust, 2 ly across, collapses under its own gravity, forming the solar nebula that will give rise to the solar system.
2. Protosun forms at the center of the solar nebula.
3. Rotation flattens the nebula, forming a disk around the protosun.
4. Condensation and accretion give origin to planetesimals.
5. Protoplanets gradually form in a rotating disk.
6. After the terrestrial protoplanets differentiate, they become full-fledged planets.
7. The Jovian planets did not differentiate. They grew big by attracting material directly from the nebula.
8. Finally, as the luminosity of Sun increased, gas and dust were eventually blown away. Left over planetesimals were swept away by gravitational interaction with planets.

OTHER SOLAR SYSTEMS

The planetary nebula that gave rise to the Sun and the planets formed as a molecular cloud collapsed. We believed that the same process has happened all throughout the galaxy and other galaxies. So we expect many stars to have planets. Current telescopes cannot directly image extrasolar planets. Planets are too small and too close to their stars to image them directly.

Planetary systems around other stars have been detected by observing the motion of the stars as the planets revolve around their center of mass. As a planet and a star orbit around their center of mass, the star wobbles slightly to and fro. This motion is very small, but it can be measured as a Doppler shift. This shift allows astronomers to measure the mass and the time the planet takes to revolve around the star.

The first extrasolar planet discovered in this way circles the star 51-Pegasi.

More than a thousand stars with planets, or with planets in formation, have been detected in the last 20 years. The planets discovered so far are closer in mass to Jupiter.

Finding an Earth-like planet won't be easy because they are small and planets are relatively close to the parent star.

In the Orion nebula, dusty disks around young protostars have been observed. In these disks, planets might be forming.

In 2004, the European Southern Observatory detected in the infrared a planet orbiting a brown dwarf.

Answers

1. d
2. a
3. c
4. a
5. c
6. c
7. d
8. d
9. c
10. c
11. A one-sixth of the escape speed (24/6 =) 4 is larger than the average speed of O on that planet, so the atmosphere contains oxygen.
12. The escape speed in the minimum initial speed that an atom, a molecule, and/or an object needs in order to escape from the surface of a star, a planet, or Moon.
13. a
14. Mercury. The low escape speed and the high temperature of Mercury have caused the majority of the gases to escape from its surface.
15. c
16. a
17. b
18. d
19. b
20. e
21. e
22. b
23. b
24. It is the time needed for half of the atoms of a radioactive sample to decay.
25. d
26. d
27. ~ 72% H, He, 1% of heavier elements
28. c
29. c
30. d
31. b
32. a
33. a
34. d
35. b
36. a
37. a
38. d
39. d, Silicates and metals condensed near the Sun and also in the outer parts of the solar nebula, where the Jovian planets formed.
40. e
41. d
42. e
43. e
44. c
45. d

UNIT 3

Now we are going to study the birth, evolution, and death of the stars. This will be done in the following three chapters.

Chapter six discusses the general properties of the stars, and chapters seven through nine discuss the evolution of the stars from birth to death.

All that we know about the stars has been extracted from the light that comes from them. The spectra of the light from the stars reveals the secrets of the stars.

The spectra are produced by the stars' atoms. We assume that the atoms behave in the same way everywhere in the entire universe. If this were not the case, it will be impossible to learn anything from the star's light.

If our knowledge of light is faulty, so is our knowledge of the stars.

"All men have stars, but they are not the same things for different people. For some, who are travelers, the stars are guides. For others, they are no more than little lights in the sky. For others, who are scholars, they are problems...."

—Antoine de Saint-Exupery, *The Little Prince*

And what are the stars for you?

CHAPTER 6

Important Properties of the Stars

The most important properties of the stars are distance, temperature, composition, luminosity, mass, and size. We are going to learn how astronomers measure and correlate these properties. The main points of this chapter are as follows:

1. Distance to stars (Parallax and Spectroscopic Method)

2. Stellar magnitude, brightness, and luminosity.

3. Surface temperature of the stars and spectral classes.

4. The Herztsprung–Russell (H–R) diagram.

5. Binary stars.

6. Mass of stars and the mass and luminosity relation.

MEASURING THE DISTANCE TO THE STARS

There are two methods of measuring the distance to the stars: **stellar parallax** and **spectroscopic parallax**.

Distance to Stars Using Stellar Parallax

The main aspects of stellar parallax were discussed in chapter one. We saw that parallax is the apparent change in the position of an object relative to some distant background due to the change in the location of the observer. We also saw that stellar parallax, **p**, is very small and has to be measured with a telescope by viewing the same group of stars 6 months apart, as shown in Figure 6-1.

We also saw that the distance, **d**, in parsecs (pc), and the parallax angle, **p**, in arc seconds, are related by the following equation

$$\text{distance (parsec)} = \frac{1}{\text{parallax (arc second)}}$$

or $\quad d = \dfrac{1}{p}$ (pc).

Q1. How far away is the star Epsilon Auriga, in the constellation Auriga, that has a parallax of 0.00143 arc seconds?

 a. 7.8 ly
 b. 7.8 pc
 c. 7,990 AU
 d. 699 pc

Q2. The star Betelgeuse, in the constellation Orion, is 130 pc away. What is its parallax in arc seconds?

 a. 0.0077
 b. 0.076
 c. 0.76
 d. 7.6

Q3. Star Susej is 10 parsec away and star Yram is 40 parsecs away. Which star has the greatest parallax angle?

 a. Susej
 b. Yram
 c. The parallax angle is the same for both stars

Q4. Which of the following star is closest to us?

 a. Procyon (parallax = 0.29 arc second)
 b. UV Ceti (parallax = 0.39 arc second)
 c. Canopus (parallax = 0.01 arc second)
 d. Altair (parallax = 0.20 arc second)

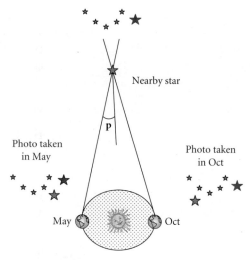

FIGURE 6-1. *When the same group of stars is viewed 6 months apart, they show a small parallax.*

Example

The Wolf 359 star has a parallax of 0.418 arc second. How far away is the star?

$$d = \frac{1}{p} \text{ pc} = \frac{1}{0.418} \text{ pc} = 2.39 \text{ pc}$$

The Wolf 359 star is the third closer star system to Earth.

Problems 2 through 5 give a good review of the concept of parallax.

With Earth-bound telescopes, the smallest parallax measured is only 0.02" arc seconds. This gives a maximum distance of 50 pc (d = 1/0.02 = 50 pc). There are only approximately 10,000 stars within this distance.

Space telescopes, located above the Earth's atmosphere, have better resolution and are able to resolve parallaxes as low as **0.01 arc seconds**. This increases the distance measurements to 100 pc = 326 ly. Even though 326 ly seems a great distance to us, this is a spec in the vastness of the universe. Only a million or so stars are located within this distance.

The Hipparcos satellite, launched in 1999, with a telescope on board measured the stellar parallax to thousands of stars.

The **spectroscopic method**, which is based on the star's brightness, or magnitude, is used to measure larger distances. Before we study this method, we need to say a few words about stars magnitudes.

Apparent and Absolute Visual Magnitude of the Stars

When we look at the sky on a clear and moonless night far away from city lights, we see countless numbers of stars. Not all of the stars have the same brightness. Some stars, like the main stars in the Orion constellation (a winter constellation), are very bright. Other stars are very faint and difficult to see. Astronomers classify the stars according to their brightness. Hipparchus of Samos (190–120 BC) was the first astronomer to do so.

He classified the stars into six different classes or visual magnitudes.

The brightest stars were class one, the second brightest were class two, and the least brightest were class six. Class six represents the dimmest stars visible with the naked eye.

The visual magnitude is the number that tells us how bright the stars are when observed from Earth. This is known as **apparent visual magnitude**.

The apparent visual magnitude, m_V, tells us how bright a star looks to us. It is called **apparent magnitude** because it depends on the distance to the stars. If you are looking at two identical stars, the closest star is the brightest star.

When the telescope was discovered, the magnitude scale was extended to include the brighter and the dimmer stars that were now visible. Brighter stars than class one have zero or negative apparent visual magnitude. The dimmer stars which are invisible to the naked eye have visual apparent magnitudes larger than six.

The Hubble Space Telescope can detect objects with an apparent visual magnitude of $m_V = 27$.

Is there a magnitude that does not depend on the distance to the stars? In other words, is there a fixed conventional distance that is used as a reference to compare the magnitude of the stars?

The answer is yes, and that is the **absolute visual magnitude**, M_V. Astronomers have selected the distance of 10 parsecs (approximately 33 ly) as a standard distance to compare the intrinsic brightness, or absolute visual magnitude, of the stars.

The apparent visual magnitude of a star measured from a distance of 10 pc is known as absolute visual magnitude (M_V).

In summary, the apparent visual magnitude, m_V, indicates how bright a star looks to an observer on Earth, and the absolute visual magnitude (M_V) how bright it looks from a distance of 10 pc. See Table 6-1.

The following example will help us to see the difference between the two magnitudes.

The apparent visual magnitude of the Sun is -26. If you go on an imaginary spacecraft and view the Sun from a distance of 10 pc the Sun would look very dim and would have an apparent visual magnitude of only 4.8. This is the absolute visual magnitude of the Sun. The Sun is not a very bright star. See Figure 6-2.

Q5. Which of the following stars is farthest from us?
a. Procyon (parallax = 0.29 arc second)
b. UV Ceti (parallax = 0.39 arc second)
c. Canopus (parallax = 0.01 arc second)
d. Altair (parallax = 0.20 arc second)

Q6. The parallax angle of the star UV Ceti is 0.39 arc second. If you were able to measure the parallax of this star from Mars, it would be _____. (Hint: Draw a diagram while keeping in mind that the orbital radius of Mars is larger than Earth's).
a. larger than 0.39 arc second
b. less than 0.39 arc second
c. 0.39 arc second
d. this cannot be determined

Q7. The apparent visual magnitude of Sirius, in the constellation Canis Major, is -1.44. The apparent visual magnitude of Deneb, in the constellation Cygnus, is $+1.25$. Which star looks brighter?
a. Sirius
b. Deneb
c. this cannot be determined

Chapter 6—*Important Properties of the Stars*

TABLE 6-1. Magnitudes and distances for some well-known stars (from the precise measurements of the Hipparcos mission).

STAR	APPARENT MAGNITUDE	DISTANCE (pc)	ABSOLUTE MAGNITUDE	LUMINOSITY SOLAR UNITS
Sun	−26	0.0000068 (1 AU)	4.5	1
Sirius	−1.44	2.63	1.45	22.5
Canopous	−0.62	96	−5.5	13,600
Betelgeuse	0.45	130	−5.1	63,000
Arcturus	−0.05	11.25	−0.31	114
Vega	0.03	7.76	0.58	50.1
Spica	0.98	80.4	−3.55	2,250
Proxima Centaury	11	1.3	15.45	5.6×10^{-6}
Barnard star	9.54	1.3	15.45	4.3×10^{-4}

Q8. Star A has an apparent visual magnitude of +8.4. This star is visible with the naked eye.

   a. True   b. False

Q9. The absolute visual magnitude of a star is the visual magnitude measured from a distance of _____.

   a. 10 AU   b. 6 ly
   c. 10,000 km   d. 10 pc
   e. 10 ly

Q10. Star A and Star B are not at the same distance from Earth. However, they have the same apparent visual magnitude. Is it possible?

   a. Yes   b. No

Q11. In your own words, what is the difference between M_V and m_V?

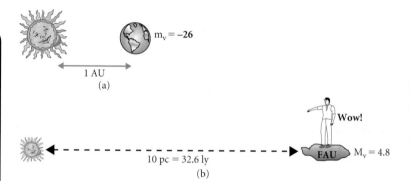

FIGURE 6-2. (a) The apparent magnitude of the Sun is $m = -26$. (b) When the Sun is viewed from a distance of 10 pc its visual magnitude is only 4.8. This is the absolute visual magnitude, M_V, of the Sun.

Spectroscopic Parallax and Distance Determination

To find the distance to the stars, the **spectroscopic parallax** method uses the values of the apparent and absolute visual magnitudes. (For simplicity, in what follows the sub index, V, will be dropped from the apparent and absolute visual magnitudes.)

It can be shown that in terms of **m** and **M**, the distance d to the star is obtained using the following equation:

$$d\,(\text{pc}) = 10^{\frac{m - M_v + 5}{5}} \quad (6\text{-}1)$$

Example

The apparent and absolute visual magnitudes of a star are $m = 3$ and $M = -2$, respectively. How far away is the star?

Solution $d = 10^{\frac{3-(-2)+5}{5}} = 10^2 = 100$ pc

LUMINOSITY AND APPARENT BRIGHTNESS

The total amount of energy that a star emits per second is known as luminosity (**L**). In chapter three, we learned that the luminosity of a star, with radius **R** and surface temperature **T**, is proportional to the square of the radius times the temperature raised to the fourth power. Replacing energy E, for the luminosity L, equation 3-7 becomes

$$L \propto R^2 T^4. \qquad (6\text{-}2)$$

The luminosity **L** is measured in Watts (1 Watt = Joules/second).

The energy, or radiation, that we received from the stars depends on the distance to the stars. This energy is known as **apparent brightness**, or **brightness, B**.

The brightness of a star is related to the apparent magnitude of that star. A five apparent visual magnitude difference corresponds to a factor of 100 in brightness. An example will help us to understand this statement.

The star, Ross-154, has an apparent magnitude of +10.22 and the star, 61-Cygni, has an apparent magnitude of +5.22. 61-Cygni, appears 100 times brighter than the star, Ross-154 because the difference between the apparent magnitudes is 10.22 − 5.22 = 5. (Remember that the scale is backwards).

The same relation holds for the absolute magnitude and the luminosity; i.e., a five absolute visual magnitude difference corresponds to a factor of 100 in luminosity.

Example

The Sun has an absolute magnitude +4.8, and the star, Lacaille 9352, has an absolute magnitude of +9.8. How many times more luminous is the Sun than Lacaille 9352?

Solution The difference between the absolute magnitudes is 9.8 − 4.8 = 5. Therefore, the Sun is 100 times more luminous than Lacaille 9352.

RELATION BETWEEN APPARENT BRIGHTNESS AND THE DISTANCE (d) TO THE STARS

If we compare two stars that have the same luminosity located at different distances, the closer star looks brighter. This is because as light moves away from a star, it spreads over larger and larger surfaces and its brightness decreases. A diagram will help us to understand the relationship between luminosity (**L**), brightness (**B**),

---

Q12. A star located 10 pc away has an apparent magnitude of −1.5. What is its absolute magnitude?

a. 2   b. 5
c. −1.5   d. +1.5

Q13. The apparent and absolute visual magnitudes of Sirius are m = −1.4 and M_V = +1.5, respectively. How far away is it?

a. 10 pc   b. 3.5 ly
c. 2.63 ly   d. 2.63 pc

Q14. What is the distance, in pc, to the star Canopus? (Hint: use Table 6-1.)

a. 10   b. 20
c. 95   d. 140

Q15. Magnitude one stars are 100 times brighter than magnitude six stars.

a. True   b. False

Q16. A difference of five apparent magnitudes equals a factor of _____ times in brightness.

a. 4   b. 10
c. 32   d. 100

Q17. The apparent magnitude of the star Spica is +0.93 and the apparent magnitude of the star Susej is +5.93. Which of the following is true?

a. Spica is 10 times brighter than Susej
b. Spica is 10 times fainter than Susej
c. Spica is 100 times brighter than Susej
d. Spica is 100 times fainter than Susej
e. Not enough information to decide

Q18. The total amount of energy emitted by a star each second best defines _____.

a. brightness
b. intensity
c. temperature
d. luminosity

Q19. The amount of energy an observer on Earth receives from a star each second best defines

a. brightness
b. intensity
c. temperature
d. luminosity

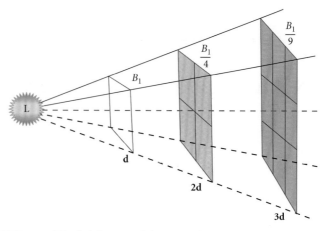

FIGURE 6-3. *The brightness of the stars obey an inverse square law.*

and distance (**d**). Assume that different observers are viewing the same star from a distance (d, 2d, 3d) and so forth, as shown in Figure 6-3. From the figure, it is clear that the energy (brightness, **B**) that each observer receives per unit surface depends on the distance to the star.

The energy received by the observer at a distance, **d**, is distributed over a surface area, **A**. The energy received by second observer at a distance 2d is distributed over a surface area four times larger than in the previous case. Therefore, the star appears to him one-fourth less bright. Similarly, to the observer located at a distance 3d, the star appears one-ninth less bright than to the first observer. To the fourth observer, the star would appear one-sixteenth less bright, and so forth.

Now, if we increase the luminosity, **L**, of the star, the brightness that each observer receives will be increased in the same proportion to which the luminosity of the star increases.

This imaginary experiment tells us that the brightness, **B**, is proportional to the luminosity, **L**, of the star, and inversely proportional to the square of the distance (**d²**), or

$$\text{Apparent brightness} \propto \frac{\text{luminosity}}{(\text{distance})^2} \tag{6-3}$$

$$B \propto \frac{L}{d^2}.$$

Example

Consider two identical stars, A and B. Each star has the same luminosity **L**, but star A is three times as distant as star B. How many times brighter is star B than star A?

Solution The first step is to draw the problem as shown in Figure 6-4. Both stars have the same luminosity **L**, so in

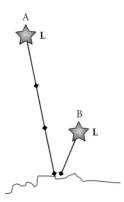

FIGURE 6-4. *The received energy, or brightness, of two identical stars depends on the square of the distance.*

equation (6-2), we can ignore the luminosity. It is enough to take the ratio of the distances and square the result. $(3/1)^2 = 9$. Since this is an inverse relation, star B is nine times brighter than star A.

Another Example

At night, you see the head lights of an approaching airplane 5 miles away. How many times brighter are the headlights when the airplane is 2.5 miles away?

Solution The square of the ratio of the distances is $(5/2.5)^2 = 4$. Since the brightness follows an inverse square relation, this tells us that when the plane is 2.5 miles away, it looks four times brighter than when it is at a distance of 5 miles.

Another Example

Two stars, X and Z, are at the same distance. However, the luminosity of star X is 100 times brighter than the luminosity of star Z. How many times brighter is star X than star Z?

Solution The distance is the same for both stars; therefore, it will not affect the brightness. The brightness increases in the same proportion to which the luminosity increases. As a result, star X looks 100 times brighter.

The Earth receives about 1,366 W/m² from the Sun. The Sun's average distance is approximately 150 million kilometer from us. Using equation 6-2 and these values, we see that the Sun emits about 4×10^{26} W or 4×10^{26} J/s.

The 1,366 W/m² value is known as the solar constant because the Sun's output is fairly constant.

Our galaxy has billions of stars with different luminosities. Some are as luminous as our Sun, others are many times more luminous, and others have only a fraction of the Sun's luminosity. The luminosity of the stars is usually expressed in terms of the luminosity of the Sun. The star, Sirius, for example, is 22 times as luminous as the Sun, and the Barnard star is only 4,000 times of the luminosity of the Sun.

Q20. The star Leonora has an apparent magnitude of +6.45 and an absolute magnitude of −6.45. If we could increase the distance to the star, its apparent magnitude would _____ and its absolute magnitude would _____.

a. increase; increase
b. increase; decrease
c. decrease; increase
d. decrease; decrease
e. increase; not change

Q21. Consider two identical stars, A and B, each with the same luminosity L. Star A is 20 ly away and star B is 5 ly away. How many times brighter is star B than star A?

a. 4   b. 8
c. 16   d. 24

Q22. Star B is three times as distant as star A. Both stars have the same luminosity. How many times brighter is A than B?

a. 9   b. 21
c. 3   d. 81

Q23. Two stars, A and B, are at the same distance from Earth and have the same absolute visual magnitude. Star A is _____.

a. is brighter than star B

b. is less bright than star B

c. is as bright as star B

Q24. If the Earth were moved twice as far from the Sun from its present location, the Sun would look.

a. twice as bright

b. twice as dim

c. four times as bright

d. four times as dim

SURFACE TEMPERATURE OF THE STARS

If you look at the sky on a clear moonless night and away from city lights, you discover that the stars have different colors. Some stars are bluish, some are orange, and many are red. Their color is an indication of their surface temperature. This is because the stars are blackbodies, and the star's surface temperature can be obtained from their continuous spectra. This was explained in chapter three.

More than 100 years ago, astronomers discovered that the strength of the absorption lines of the stellar spectra reveals the surface temperature of the stars. Astronomers used the stellar absorption to measure the surface temperature of the stars.

If the surface temperature of the stars is between 12,000 and 40,000 degrees, most of the hydrogen atoms are ionized. Thus, the hydrogen absorption lines, or Balmer lines, in the spectra are very weak. However, the spectra shows lines of neutral helium.

For surface temperatures between 5,000 and 11,500, the electrons of the hydrogen atom are in orbit number two and is able to absorb the corresponding photons to produce strong Balmer absorption lines. For lower surface temperatures, the photons do not have enough energy to propel the electrons of hydrogen to higher orbits. As a result, the spectra has very weak Balmer absorption lines. At these low temperatures, the absorption lines of other elements are fortunately present.

By analyzing the strength of the absorption lines, as shown in Figure 6-5, astronomers have classified the stars into seven traditional spectral classes: O, B, A, F, G, K, and M. The stars with the highest surface temperatures are the O stars, and the stars with the coolest surface temperatures are the M stars.

The spectra is usually displayed with a diagram, wavelength vs intensity, as shown in Figure 6-6. As we saw earlier, the absorption lines in an intensity wavelength diagram appear as dips below the continuous curve. Recall that the Balmer absorption

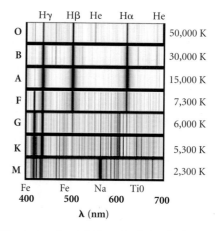

FIGURE 6-5. *The strength and the number of absorption lines of the star's spectra depends on their surface temperature.*

lines have wavelengths of Hγ = 434 nm, Hβ = 486.1 nm, and Hα = 656.3 nm.

Table 6-2 gives the spectral classification and the approximate temperature range for each spectral class. The table includes two new classes of cooler stars that have been discovered since 1998: the L and T dwarfs.

Within each spectral class, there are 10 subtypes from 0 through 9. For example, the A class has the following subclasses: A0, A1, A2, ... A9. An A2 star is hotter than an A6. There is no A10.

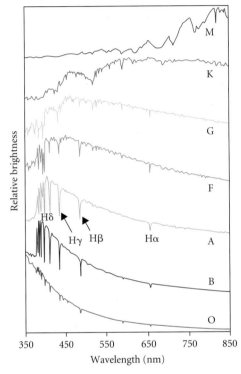

FIGURE 6-6. *The hydrogen lines are weaker for stars with high and low surface temperatures.*

TABLE 6-2. *Spectral classification of the stars according to the surface temperature.*

SPECTRAL CLASS	TEMPERATURE RANGE (K)	COLOR
O	28,000–50,000	Blue
B	12,000–28,000	Bluish white
A	8,000–12,000	White
F	6,500–8,000	Yellowish white
G	5,000–6,500	Yellowish
K	3,500–5,000	Orange
M	2,500–3,500	Reddish
L	1,200–2,200	Infrared
T	Under 1,200	Infrared

Q25. The spectral types of the star Capella and of the Sun are G1 and G2, respectively. It follows that

a. Capella is hotter than the Sun
b. Capella is as hot as the Sun
c. the surface temperature of Capella is twice the Sun's surface temperature

Q26. The temperature of star X is four times the temperature of star Y, other conditions being equal. This tells us that the luminosity _____.

a. of star X is four times the luminosity of star Y
b. of star Y is four times the luminosity of star X
c. of star X is 16 times the luminosity of star Y
d. of star Y is 16 times the luminosity of star X
e. of star X is 256 times the luminosity of star Y

Q27. The radius of star X is four times the radius of star Y, other conditions being equal. This tells us that the luminosity _____.

a. of star X is four times the luminosity of star Y
b. of star Y is four times the luminosity of star X
c. of star X is 16 times the luminosity of star Y
d. of star Y is 16 times the luminosity of star X
e. of star X is 256 times the luminosity of star Y

CHEMICAL COMPOSITION OF STARS

A star's composition is another piece of information that we obtain from the stars' spectra. This information is given by the position and intensity of the absorption lines.

For example, if the spectrum of a star contains lines corresponding to the element calcium, we conclude that the star must contain calcium.

The chemical composition of the stars is determined by comparing the absorption spectra of the stars with the spectra of the elements that we have stored on our computers.

The results show that the composition of the stars is similar to the Sun's composition:

71% of hydrogen by mass (91% atoms)

27% helium by mass (8.9% atoms)

2% nitrogen, carbon, metals, and other elements (1.1% by mass).

The universe is roughly three-quarters hydrogen and one-quarter helium, with 1 or 2% of other elements.

THE HERTZSPRUNG AND RUSSELL (H–R) DIAGRAM

Astronomers have measured the luminosity and the surface temperatures of thousands of stars. To analyze these large data, astronomers typically use a diagram, plotting the luminosity of the stars as a function of temperature, as shown in Figure 6-7. This type of diagram is known as the Hertzsprung–Russell diagram, or a simple H–R diagram.

Usually the luminosity of the stars is given in solar luminosities, or solar units, and the temperature is given in Kelvin. Often the temperature is indicated with the spectral classes (O, B, etc.). Since the luminosity (L) and the absolute visual magnitude (M_V) of the stars are related, the H–R diagram usually includes this relation explicitly, displaying in one vertical axis the luminosity in solar units and in the other axis the absolute magnitude. For example, in Figure 6-7, the star S, the Sun, has a luminosity of 1 solar unit and an absolute magnitude of five. The star P has an absolute magnitude of zero and 100 solar luminosities.

The luminosity and temperature of the stars covers a large range of values, so, logarithmic scales, instead of linear scales are used. In the log scale, the partitions are equal but they change by a factor of 10. See Figure 6-8.

When the luminosity of the stars is plotted on an H–R diagram, they fall into any of the following groups: main sequence stars, super giants, giants, and white dwarfs, as shown in Figure 6-7.

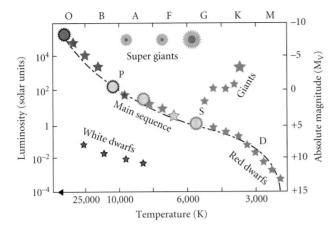

FIGURE 6-7. *H–R diagram. The vertical and horizontal axis are logarithmic scales.*

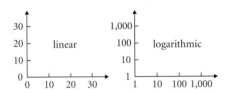

FIGURE 6-8. *Comparing a linear scale with a logarithmic scale. In the log scale, the partitions are equal but they change by a factor of 10.*

Figure 6-7 shows that the luminosity and the absolute magnitude of the stars are related. What is the nature of this relation?

As mentioned previously, a five absolute visual magnitude difference corresponds to a factor of 100 in luminosity. Keep in mind, however, that the more luminous star has the smaller absolute magnitude. If we have two stars, a and b, with luminosities, **La** and **Lb**, and absolute magnitudes, **Ma** and **Mb**, this relation is expressed with the following equation:

$$La = 2.51^{(Mb - Ma)} Lb.$$

To compare the luminosity of the stars with the luminosity of the Sun, we assume that star b is the Sun in the previous equation. Next, replace **Lb** by 1 solar unit and, approximating the absolute magnitude of the Sun from 4.8 to 5, the luminosity of the stars expressed in solar luminosity is given by

$$L = 2.512^{(5 - Ma)} \text{ solar luminosities.} \tag{6-4}$$

Example

What is the luminosity of a star if its absolute magnitude is zero?

$$L = 2.512^{(5 - 0)} \sim 100 \text{ solar luminosities}$$

See Figure 6-7.

Use Figure 6-19 to answer questions 28 through 31.

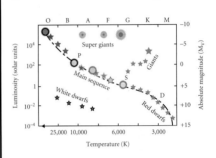

FIGURE 6-19. *H–R diagram. The vertical and horizontal axis are logarithmic scales.*

Q28. What is the luminosity, in solar units, of a star whose absolute magnitude is −5?

a. 10   b. 100
c. 1,000   d. 10,000

Q29. A star with a surface temperature of 3,000 K and 100 solar luminosity is a _____.

a. white dwarf
b. red dwarf
c. main sequence
d. red giant

Q30. A 100 solar luminosity star has an absolute visual magnitude of _____.

a. −5   b. 0
c. +5   d. +10

Chapter 6—*Important Properties of the Stars*

Main Sequence (MS) Stars

The main sequence group of stars runs along the diagonal of the H–R diagram. See Figure 6-9.

Notice that main sequence stars with high surface temperatures are very luminous. Conversely, low surface temperature stars have low luminosity. These stars are called **red dwarfs**.

Some important properties of main sequence stars are as follows:

1. They are the most common type of star. Approximately 90% of the stars in the Milky Way are main sequence. Our Sun is a main sequence star.
2. They produce energy by fusing hydrogen into helium in or near their cores.
3. They are in hydrostatic equilibrium. (The outward force of pressure is balanced by the inward force of gravity.)
4. Their luminosity, size, temperature, and mass do not change much.
5. They range from highly luminous hot O stars with more than 10,000 times the luminosity of the Sun (upper left corner of diagram), to the low luminous and cool red dwarfs (lower right corner) with less than one-thousandth of the luminosity of the Sun.
6. Stars spend about 90% of their lives as main sequence.

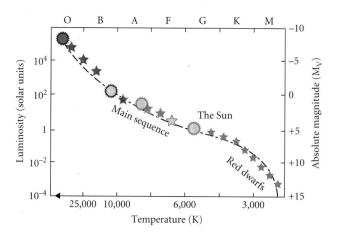

FIGURE 6-9. *Main sequence stars.*

Giant Stars

These stars are evolved, postmain sequence stars. There are several subtypes, consisting of the super giants, red subgiants, and the red giants. The super giants are located on the top of the diagram, and the subgiants and red giants are located toward the right corner above the main sequence stars. See Figure 6-10.

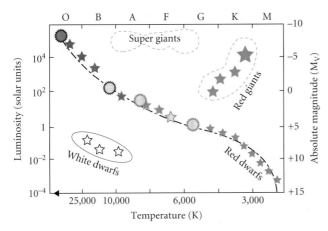

FIGURE 6-10. *Main sequence stars and RED GIANTS.*

Both of these groups of stars were once main sequence stars. These stars have the following characteristics:

1. They are evolved postmain sequence stars.
2. They are very big, very luminous, and cool stars.
3. They produce energy by fusing heavier elements than hydrogen.
4. They are very old stars which are close to the moment of death.

Red giants are cool and very luminous. Therefore, equation

$$L \propto r^2 T^4 \qquad (6\text{-}5)$$

tells us that they have to be big. Why?

White Dwarfs

White dwarfs have the following characteristics:

1. They are also postmain sequence stars, located in the left bottom corner of the diagram.
2. They are hot and not too luminous. Therefore, according to equation 6-3, they have to have a small radii.

Let's apply equation 6-2 in the solution of the following examples.

Example 1

The red giant, Betelgeuse, located in the constellation Orion, has a surface temperature of 3,500 K, and a radius 630 times the radius of the Sun. How many times more luminous is Betelgeuse than the Sun?

Solution We know that the luminosity for both stars is given by equation 6-2. Let's take the ratio of their luminosities:

$$\frac{L\ \text{star}}{L\ \text{Sun}} = \left(\frac{3{,}500}{5{,}800}\right)^4 \left(\frac{630}{1}\right)^2 = 52{,}631.$$

Q31. Which of the following groups of stars is the most luminous:

a. white dwarfs
b. red dwarf
c. red giants
d. super giants

Q32. Stars A and B have the same radii, but the surface temperature of star B is 2.5 times the surface temperature of star A. How many times more luminous is star B than A?

a. 6.25 b. 15.6
c. 39.1 d. 40.5

Q33. Stars A and B have the same radii, but the surface temperature of star A is 9,350 K and the surface temperature of star B 15,000 K. How many times more luminous is star B than A?

a. 2 b. 4
c. 6.6 d. 16

Use **Figure 6-20** to answer questions 34 through 39. Note that there is a relationship between the luminosity and the absolute magnitude of the stars. (See right vertical axis.)

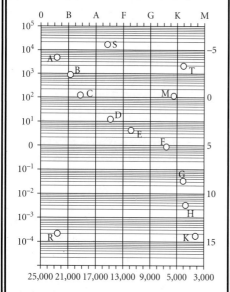

FIGURE 6-20.

Q34. Label and give the units of the four axes.

Q35. A star like the Sun is star _____. Give the reason for your answer.

Q36. Which of the following characteristics do stars G, F, H, and A all have in common?

    a. All of them are red giants

    b. All of them are main sequence stars

    c. All of them are red dwarfs

    d. All of them are white dwarfs

Thus, Betelgeuse is about 53,000 times more luminous than the Sun.

In practice, you can do the following: Take the ratio of the two temperatures. Then raise it to the fourth power. Next, take the ratio of the two radii and square it. Finally, do the product of these two numbers.

Example 2

The surface temperature of the star, Canopus, located in the constellation Carina, is 2.6 times the surface temperature of the Sun. Its radius is 65 times the radius of the Sun. How many times more luminous than the Sun is Canopus?

Solution The problem already gives the radii and temperature ratios, so all you have to do is square the radii ratio and raise to the fourth power the temperature ratio. Finally, multiply these two numbers as follows:

$$(65)^2 \times (2.6)^4 = 19{,}307.$$

Canopus is 19,307 times more luminous than the Sun.

Example 3

Consider a star as big as the Sun, having the same radius but twice as hot. How many times more luminous is the star than the Sun?

Solution Both have the same radii. Therefore, all we have to do is take the ratio of the temperatures of the stars and raise it to the fourth power

$$(2)^4 = 16 \qquad \text{So,} \qquad L_{star} = 16\, L_{Sun}.$$

Example 4

In Figure 6-11, the stars Z and T are on the same horizontal line. Which star is bigger?

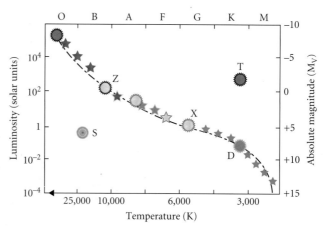

FIGURE 6-11. *H–R diagram for examples 4 and 5.*

Solution Both stars have the same luminosity, but star Z is hotter than star T. Therefore, equation 6-2 tell us that star R has to be larger. The logic behind this reasoning is the following. If two stars have the same luminosity but different sizes and different temperatures, the hotter star has to be smaller.

Example 5

Again in Figure 6-11, consider the main sequence stars Z and D. Decide which star is bigger.

Solution Star Z is hotter and more luminous than star D. Therefore, equation $L \propto R^2 T^4$ tell us that D has to be smaller.

Example 6

Stars T and D have the same temperatures. Which one is larger?

Solution Both stars have the same temperature but star T is more luminous than star D. Therefore, equation 6-2 tells us that D has to be larger. Why?

Summary

Equation 6-2 tells us that low surface temperature stars with big radii can be very luminous (red giants). Further, stars with high surface temperatures but small radii is not very luminous (white dwarfs).

LUMINOSITY CLASSES OF THE STARS

We have already classified the stars according to their surface temperature. To complete the classification, the luminosity of the star must also be considered.

The largest and most luminous stars are the super giants. They are big because their atmosphere expanded at the end of the hydrogen cycle.

The densities of the atmospheres of the giant stars are lower than their densities when they were in the main sequence phase. The H–R diagram of Figure 6-12 shows the different luminosity classes of the stars.

The width of the absorption lines depends on the density of the star's atmosphere. If the star's atmosphere is dense, like a main sequence, their hydrogen atoms have frequent collisions and the absorption lines are broad. The lesser the density, the finer the lines. This implies that the absorption lines of main sequence stars are broader than the absorption lines of the more luminous giants and super giants stars, as shown in Figure 6-12. The width of the absorption lines is used to classify the stars according to their luminescence into five luminosity classes. See Figure 6-13 and Table 6-4.

Q37. Considering the stars C and M, which star is larger? Explain.

Q38. Considering stars G and T, which star is bigger?
a. T is bigger
b. G is bigger
c. T is as big as G
d. Impossible to tell

Q39. Consider stars A and G. Which of the following is true?
a. Stars A and G are main sequence stars
b. Star G is cooler
c. Both stars are fusing hydrogen into helium near the core
d. All of the above are true

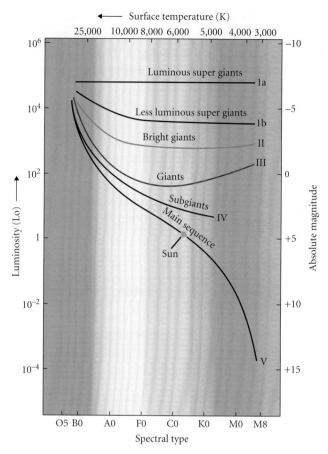

FIGURE 6-12. *Distribution of the luminosity of the stars on an H–R diagram.*

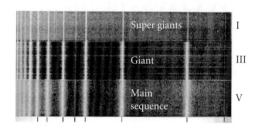

FIGURE 6-13. *The width of the absorption lines of the stars is related to their luminosity.*

TABLE 6-3. *Luminosity classes of stars. (The classes are given by roman numbers).*

Class I:	Super giants—brightest class.
Class II:	Bright giants—intermediate class, between giants and super giants.
Class III:	Giants.
Class IV:	Subgiants—faint giants, intermediate.
Class V:	Main sequence stars—faintest class.

There is no luminosity class for white dwarfs since they are not very luminous and are very difficult to see.

TABLE 6-4. *Spectral type of well-known stars.*

STAR	APPARENT MAGNITUDE	ABSOLUTE MAGNITUDE	SPECTRAL TYPE	LUMINOSITY SOLAR UNITS
Sun	−26	4.5	G2V	1
Sirius	−1.44	1.45	A1V	22.5
Capella	0.247	−0.48	G8III	180
Canopous	−0.62	−5.5	A0I	13,600
Betelgeuse	0.45	−5.1	M2I	63,000
Arcturus	−0.05	−0.31	K2III	114
Vega	0.03	0.58	A0V	50.1
Spica	0.98	−3.55	B1V	2,250
Proxima Centauri	11	15.45	K05	5.6×10^{-6}
Barnard star	9.54	15.45	M5V	4.3×10^{-4}

A complete description of a star's place in the H–R diagram includes the spectral class (temperature) and the luminosity class.

Examples

The Sun is a G2V. The G2 gives the spectral class of the Sun indicating that it has a temperature of approximately 5,800 K, and the V gives the luminosity class, which tells us that the Sun is a main sequence star.

The star, Vega (in the constellation of Lyra) is an A0V star. The spectral class is A0, indicating a surface temperature of 10,000 K, and the luminosity class is V, indicating that it is a main sequence star.

Binary Stars and Mass Determination of Stars

A binary star system consists of two stars orbiting each other around a common center of mass.

In simple terms, the center of mass, CM, is the balance point of the system where the force of gravity of star A matches the force of gravity of star B. The center of mass of the two identical stars is exactly half-way between the two stars. If the stars have different masses, the center of mass is always closer to the more massive star. See Figure 6-14.

Approximately, 75% of the stars are part of a binary system. Mass is the amount of matter in a body. There is no direct way to measure the mass of a single star—we can only measure the mass of stars that are in binary star systems.

> **Use Table 6-4 to answer questions 40 through 45**
>
> **Q40. Which two stars are the hottest?**
>
> **Q41. Which two stars are the most luminous?**
>
> **Q42. Of the five main sequence stars, which one has the lowest surface temperature?**
>
> **Q43. Which star appears the faintest to an observer on Earth?**

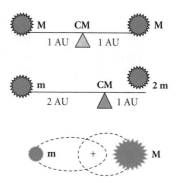

FIGURE 6-14. *The two stars in a binary system revolve around their common center of mass.*

Mass Determination of Binary Stars

The stars of a binary system always revolve around their common center of mass with the same period. Using Newton's laws of motion, we can show that the period **P** of the stars and the average separation **a** between two binary stars are related by an equation similar to Kepler's third law. This is given by

$$P^2 = \frac{a^3}{M + m}. \qquad (6\text{-}6)$$

This equation is sometimes called the **modified Kepler's third law**. In equation 6-6, the masses **M** and **m** are expressed in solar units, the period **P** in years and the average separation **a** in AU.

Notice that equation 6-6 gives the combined mass of the two stars.

Example

Two stars orbit each other with a period **P** of 10 years and an average separation **a** of 14 AU. What is their combined mass in solar masses?

$$m_a + m_b = \frac{a^3}{P^2} = \frac{14^3}{10^2} = \frac{2744}{100} = 27.44.$$

The combined mass of the two stars is 27.44 solar masses.

There are also **multiple star systems**, which are systems with more than two stars.

Types of Binary Stars

There are three different types of binary stars: **visual, spectroscopic, and eclipsing binaries**.

Visual Binary Stars

Not all binary star systems can be observed as two distinct stars with optical telescopes. The systems that appear as two different stars are called visual binaries.

Q44. Two binary stars have an average separation of 1.5 AU and a period of 1.5 years. What is their combined mass, in solar units?

- a. 1.5
- b. 2.25
- c. 4.5
- d. 6.25

Q45. In 1980, the star A of a visual binary was observed to the left of star B. In 1990, star B was to the left of star A. In 2000, they were back to the positions seen in 1980. What is the orbital period of the binary?

- a. 5 years
- b. 10 years
- c. 15 years
- d. 20 years
- e. this cannot be determined from the information given

By observing the motion of the stars for a few decades, and by using the laws of physics, their average separation **a**, and period **P**, can be predicted with confidence. These stars have long orbital periods, between 100 and 1,000 years.

Spectroscopic Binary Stars

The majority of binary stars are too far away and therefore, the telescopes cannot resolve them. Their period and average separation are calculated observing the Doppler shift of the absorption lines as the stars move back and forth around the center of mass, as shown in Figure 6-15.

Eclipsing Binary Stars

If the plane of the binary system is nearly edge-on to our line of sight, then when one of the stars passes in front of the other an eclipse happens. This produces a variation in the amount of light that we receive from the stars as shown in the light curve diagram of Figure 6-16. Both the visual and the spectroscopic binaries can be observed as eclipsing binaries.

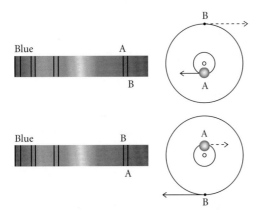

FIGURE 6-15. *The motion of spectroscopic binaries are observed by means of the Doppler effect.*

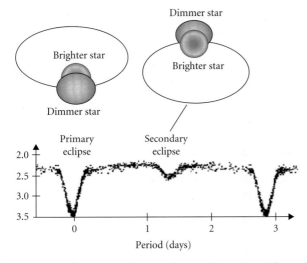

FIGURE 6-16. *Light curve of two eclipsing binaries. When the stars eclipse each other, the intensity of the light curve decreases.*

The light curve of an eclipsing binary gives information about the radii, the period, and the average separation of the two stars. This information is used to determine the mass of the two star by means of equation 6-6.

Figure 6-16 shows the light curve of the Algol system, located in the constellation Perseus. The blue star is the brighter and hotter Algol A, and the yellow star is the cooler and dimmer Algol B. The system orbits the common center of mass every 2.85 days. The primary eclipse is where the dimmer star is in front of the brighter blue star, blocking large quantities of light from the brighter star. The light curve shows a large dip because the dimmer star is blocking large quantities of light. The secondary eclipse occurs when the brighter hotter star is in front of the dimmer, yellow star. In this case, the system does not lose much light and the dip in the light curve is smaller.

Figure 6-16 shows that when we see both stars, the apparent magnitude is 2.1. However, during the primary eclipse it dips to 3.4. This is an example of a partial eclipsing binary because the two stars are not completely aligned with the Earth. This eclipse lasts about 10 h.

The epsilon Auriga, located in the constellation Auriga, represents a long period binary system. The primary eclipse occurs every 27.1 years and lasts approximately 2 years. Because the eclipse is so long-lasting, the eclipsing body must be gigantic.

We have seen that by measuring the average separation and the period of binary stars, we can calculate their combined mass.

Further observations allow us to obtain the mass ratio of the two binaries, m_1/m_2. By combining these two results, the individual mass of the stars can be calculated.

In this way, the mass of thousands of main sequence stars has been determined. When the luminosity **L** (in solar units) of these stars is plotted as a function of the mass **M** (in solar masses) on a logarithmic graph, most of the stars approximately fall along a straight line, as shown Figure 6-17. Mathematically, this situation can approximately be described by the following power law.

$$\mathbf{L} = \mathbf{M}^{3.5}. \tag{6-7}$$

This expression allows us to determine the luminosity of a main sequence star if we know its mass. Conversely, if we know the mass of the main sequence star, we can determine its mass.

Example

What is the luminosity, in solar units, of a main sequence star of 2 solar masses?

Solution $\mathbf{L} = (2)^{3.5} = 11.3$ **solar units.**

This star is 11.3 times more luminous than the Sun. Verify this result with the diagram of Figure 6-17.

Q46. A 5 solar mass main sequence star has a luminosity of about _____ solar luminosities:

a. 1,000   b. 2,140

c. 3,200   d. 280

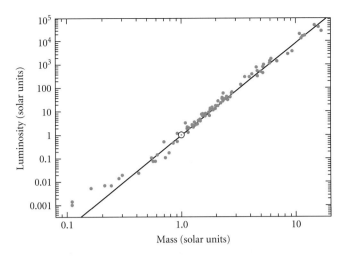

FIGURE 6-17. *Mass luminosity relationship for main sequence stars.*

Credit: Adapted from data compiled by Prof R. Pogge, Ohio State University

Example

What is the mass of a main sequence star of 1,000 solar luminosities?

Solution Solving for m in equation 6-5, we derive

$$M = \sqrt[3.5]{L}.$$

Replacing the numbers, we find

$$M = \sqrt[3.5]{1,000} \text{ solar mass} = 7.2 \text{ solar mass}.$$

The so-called **Russell–Vogt theorem** argues that all the properties of the stars are primarily determined by their mass. Considering this, it is no wonder why mass is the most important parameter of the stars.

If the luminosity and the surface temperature of main sequence stars is known, equation 6-2, gives the radii of these stars. This relation, shown in Figure 6-18, indicates high mass main sequence stars are bigger than lower mass main sequence stars.

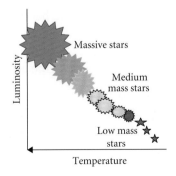

FIGURE 6-18. *The mass of main sequence stars follows a mass luminosity relation.*

> **Q47.** An 0.8 solar mass main sequence star has a luminosity of about _____ solar luminosities:
>
> a. 2   b. 1.2
>
> c. 0.8   d. 0.5
>
> **Q48.** On an H–R diagram, a main sequence star of 25 solar mass would be located in the _____.
>
> a. middle of the diagonal
>
> b. upper left corner
>
> c. lower right corner
>
> d. lower left corner

Chapter 6—*Important Properties of the Stars*

Answers

1. d
2. a
3. the closer star has the largest parallax. To prove, draw a diagram similar to the one in Figure 5-1.
4. b
5. c, The smaller the parallax, the more distance the star.
6. a
7. a
8. b
9. d
10. a, The more luminous is more distant.
11. The absolute magnitude tells us how bright a star looks from a distance of 10 pc, and the apparent magnitude tell us how bright it looks to us independent of its distance.
12. c, The absolute magnitude is measured from a distance of 10 pc.
13. d
14. c
15. a
16. d
17. c, The difference in magnitude between the two stars is: 5.93 − 0.93 = 5. A five apparent magnitude difference corresponds to a factor of 100 in brightness.
18. d
19. b
20. e, The apparent magnitude depends on the distance and the absolute magnitude does not depend on the distance.
21. c
22. a
23. c
24. d
25. a
26. e
27. d
28. d
29. d
30. b
31. d
32. c
33. c
34. Temperature in K (bottom horizontal axis), Spectral class (top horizontal axis), Luminosity in solar units (left vertical axis), Absolute magnitude (right vertical axis).
35. F, Because it has 1 solar luminosity and a temperature of 5,800 K.
36. b
37. Star M, According to the Stefan Boltzmann law, $L \propto R^2 T^4$, if two stars have the same luminosities but different temperatures, the cooler star is the larger.
38. a, Using $L \propto R^2 T^4$, we see that if two stars have the same temperatures and different luminosities, the largest is the more luminous.
39. d

40. Canoupus, Vega
41. Canopus and Betelgeuse
42. Barnard star
43. Proxima Centauri
44. a
45. d
46. d
47. d
48. b

CHAPTER 7

The Lives of the Stars: From Birth to Main Sequence Stars

The stars, like the Sun, form from molecular clouds. We studied, in the second part of chapter five, the process that led to the formation of the Sun. The Sun is an average star, so we believe that the same mechanism that gave rise to the Sun is also responsible for the formation of the stars in our galaxy and other galaxies. Some of the concepts explained in this chapter were briefly mentioned when we discussed about the formation of the solar system. Within the last 50 years, astronomers have gained a better understanding of the process by which the stars form. In this chapter, we will analyze the most probable mechanisms that lead to the formation of the stars.

THE INTERSTELLAR MEDIUM

The matter that fills the space between the stars is called **interstellar medium** (IM). The IM is the birthplace of the stars. In some regions of space, the IM is evident, as large accumulations of matter called **nebulae**. The regions of space that at first glance seem to be devoid of matter also contain IM.

Ninety-nine percent of the IM consists of gas and 1% consists of dust.

The interstellar gas is very dilute and has a density of about 7 atoms/cm^3, for comparison, 1 cm^3 of air in a room has about 10^{20} molecules.

The composition of the gas (by mass) is about 72% of hydrogen, 26% of helium, and 2% of heavier elements than helium. If we were to count the number of atoms, it would be about 90% of hydrogen, 9% of helium, and 1% of other elements.

How do we know that the space between the stars is not empty?

When the absorption spectra of distant hot stars are taken besides the broad absorption lines of the stars, very narrow absorption lines of calcium are observed. This is shown in Figure 7-1. Astronomers know that calcium is destroyed at high temperatures and cannot be present in hot stars. They conclude from this that these narrow lines are produced as the light travels through the IM.

Further, the IM reveals its presence by the 21-cm radio radiation emitted by the atomic hydrogen present in the IM. (Later on in this chapter we will discover that the spiral arms of the Milky Way are full of cold atomic hydrogen).

Q1. What are the two main components of the interstellar matter?

a. Hydrogen and oxygen
b. Dust and ices
c. Gas and dust
d. Helium and ions

Q2. The interstellar dust probably formed _____.

a. in the atmosphere of very evolved old stars
b. in the planets of the solar system
c. in the Sahara desert
d. from the interstellar gas

Q3. Which of the following is not evidence of the existence of an IM?

a. extinction
b. narrow absorption lines in the spectra of stars
c. reddening
d. molecular bands in the spectra of cool stars

Q4. The sky appears blue during the day because _____.

a. small dust particles in the atmosphere absorb blue light less efficiently than red light
b. small dust particles in the atmosphere scatter blue light more efficiently than red light
c. the atmosphere contains a large amount of water vapor which, in large quantities, appears blue
d. 70% of Earth's surface is water and its blue color is reflected off small particles in the atmosphere.
e. the ozone in the atmosphere absorbs UV light and passes more red light than blue

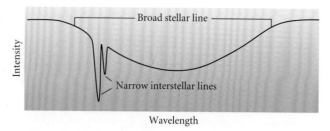

FIGURE 7-1. *The narrow bands in this spectrum are produced by the interstellar medium.*

Close to the stars, the IM is hot, with a temperature of about 10,000 K. It usually glows in the visible part of the spectrum, which produces the emission nebulae. Far away from the stars, the IM is cold and has a temperature of about 10–50 K.

The dust component of the IM is made up of small solid grains with an average size of 10^{-6} m, comparable in size to the wavelength of visible light.

The dust particles probably formed in the atmosphere of dying stars and grew to the present size by accumulating atoms and molecules from the interstellar gas. The dusty grains consist of carbon (graphite), silicates (similar to sand in composition), and metals. Each grain is probably covered by ice and water.

The gas of the IM does not affect the star light much, except for the presence of narrow absorption lines. See Figure 7-1. However, dust particles dim the light from the stars.

The darkening of star light by interstellar dust is called **interstellar extinction**. A dark nebulae is produced when dust blocks the star light.

The light from the stars encounters many dust particles as it travels to Earth. The dust particles absorb and efficiently scatter the wavelength of light, which is comparable in length to the size of the grain of dust. The long wavelengths of infrared and radio radiation are unaffected by the presence of dust in the interstellar dust. The shorter wavelengths of blue and ultraviolet (UV) light are absorbed and scattered more than any other color of the visible light spectrum.

The light we receive from the stars has less red color than when it left the star because the interstellar dust scatters and absorbs more efficiently the blue light from the visible light. See Figure 7-2. (Note that when you take blue light from white light, the visible light looks redder. Sunsets look red because the dust in the air scatters the blue light more efficiently than the longer wavelengths.)

If the stars look redder than they really are, they also appear to be cooler. For example, a star with a surface temperature of 25,000 K might appear to us as having a surface temperature of only 4,500 K, and looks redder than it actually is.

When astronomers analyze the light from the stars and galaxies, they have to make corrections for the light extinction caused by the IM.

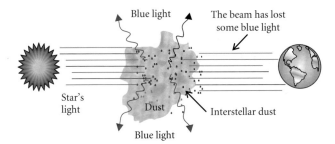

FIGURE 7-2. *The interstellar dust causes the star to appear redder than they really are.*

Interstellar dust makes background stars appear redder and dimmer.

The regions in the sky where the interstellar matter is visible form a nebula.

There are several types of nebulae. The most prominent are: **emission nebula, reflection nebula, dark nebulae, planetary nebula, and supernova remnants.**

The planetary nebulae and the supernova remnants are produced when the stars die, so we will postpone their discussion until we study the death of the stars.

Emission Nebula or HII Regions

The **emission nebulae** are diffused clouds of ionized hydrogen gas and dust. They are also called **HII regions** because the hydrogen atoms in the nebulae are ionized. Regions containing neutral hydrogen are known as **HI regions**.

The dust in the nebula is distributed over the entire body of the nebula. In some regions, the dust is visible as dark intertwined vanes.

The emission nebulae contain young and hot O and B stars that emit large quantities of UV radiation. The radiation is absorbed by the hydrogen atoms and becomes ionized. Recall that when the electron in the hydrogen atom absorbs enough energy, it leaves the hydrogen atom ionized. When the free electrons recombine with the hydrogen nuclei, the electrons fall to lower energy levels, emitting the familiar red, blue, purple, and violet Balmer hydrogen emission lines. The blending of these colors gives the characteristic red-pink color of the emission nebulae. The red Hα radiation usually dominates and the red color prevails.

The previous discussion tells us that an emission nebulae, or HII region, can be identified by its emission line spectrum.

The composition of the emission nebula is similar to the composition of the Sun and the molecular clouds in which they are imbedded. HII regions are usually found on the edges of cold molecular clouds. The temperature in an HII regions is about 10,000 K, while the temperature in the surrounding cloud is less than 100 K. Because the emission nebula contains hot stars, some parts of the nebula have temperatures of a few thousand degrees, 7,000–10,000 K.

Q5. The interstellar dust is transparent to _____ and _____ radiation.
 a. visible; UV
 b. gamma; visible
 c. radio; infrared
 d. microwave; X-ray

Q6. Interstellar reddening of starlight occurs because the interstellar dust _____ than any other color.
 a. scatters more efficiently red light
 b. scatters more efficiently blue light
 c. absorbs more efficiently green light
 d. absorbs more efficiently red and blue lights
 e. a and c

Q7. HII regions are regions _____.
 a. of ionized hydrogen gas
 b. where stars are forming
 c. that emit a reddish radiation
 d. all the above

Q8. Emission nebulae contain young hot stars that emit _____ radiation that ionizes the _____ hydrogen atoms of the nebula.

> **Q9. The gas of the emission nebulae is not blown away by the radiation of the stars because _____.**
>
> a. the helium atoms are heavy
>
> b. the stars inside the nebulae are very cold
>
> c. they are resilient to heat
>
> d. the dust grains absorb part of the radiation emitted by the stars

In summary, emission nebulae, or HII regions, are

1. regions of ionized hydrogen
2. regions that produce a hydrogen emission spectrum
3. regions that contain young hot O and B stars
4. sites of star formation

The dust in the nebulae and molecular clouds absorbs the UV radiation emitted by the star inside the nebulae. In the absence of dust, the UV radiation emitted by the stars would boil away the gas, destroying the nebulae and molecular clouds, after a few stars are formed.

The heat absorbed by the dust grains does not increase the temperature of the grains enough for them to emit light in the visible part of the spectrum. However, there is no doubt that the dust in the nebulae emits infrared radiation. Further, the dust grains in the molecular clouds provide the seeds around which particles condense.

Reflection Nebulae

The dust particles in the IM often scatter the blue light from nearby stars in the direction of the Earth. These type of nebulae are called **reflection nebulae**. The blue light is scattered more efficiently than any other color because, as explained earlier, the dust particles are similar in size to the wavelength of the blue star light. See Figure 7-3.

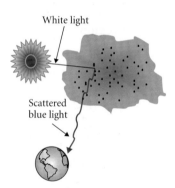

FIGURE 7-3. *Sketch of a reflection nebula.*

Our sky looks blue during the day because the molecules of nitrogen and oxygen scatter the blue light more efficiently than any other color of the Sun's light.

Dark Nebula

Within the galaxy lie vast regions of dust that obstruct the view of stars located behind these dusty regions. These regions are called **dark nebulae**. Dark nebulae are not empty regions in space. When these nebulae are observed with radio or infrared telescopes, astronomers can see the stars inside or behind the nebulae. See Figure 7-4.

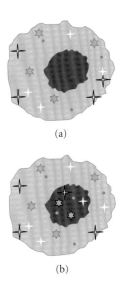

FIGURE 7-4. (a) *Dark nebula observed in visible* (b) *dark nebula observed in infrared.*

The three types of nebulae, just described, can usually be observed in the same region located about 1,500 ly from Earth.

Detection of Neutral Atomic Hydrogen or HI Regions

As stated earlier, the space between the stars is filled with interstellar matter, which contains dust and gas. The presence of dust is revealed through the extinction effect.

The presence of atomic hydrogen in the IM cannot be detected easily because its temperature is only 50 K or less.

In these conditions, the atomic hydrogen is in the ground state and does not emit visible radiation.

Fortunately, the neutral atomic hydrogen emits 21-cm wavelengths. These wavelengths are detected with radio telescopes.

The radiation is emitted when the electron in a neutral hydrogen atom undergoes a spin flip. This action causes the electron to go from a higher energy state to a lower one. When the spin axis of the proton and electron in a hydrogen atom is in the same direction, the atom has slightly more energy than when the two spins have opposite directions. See Figure 7-5.

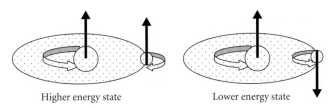

FIGURE 7-5. *As the electron of a hydrogen atom undergoes a spin flip, it emits radio radiation.*

Chapter 7—The Lives of the Stars: From Birth to Main Sequence Stars

> **Q10. The 21-cm radio radiation is emitted by _____.**
>
> a. molecules of carbon monoxide
> b. molecules of hydrogen
> c. cold atomic hydrogen
> d. ionized hydrogen
>
> **Q11. An HII region _____.**
>
> a. is rich in ionized hydrogen
> b. is identified from its emission lines
> c. contains only neutral hydrogen atoms
> d. reflects mainly the blue
> e. a and b
>
> **Q12. A reflection nebula is an interstellar dusty cloud that _____.**
>
> a. scatters mainly blue light toward us from stars that are close by
> b. contains young hot stars that warm up the dust grains causing it to emit blue light
> c. scatters the red light and absorbs the blue light emitted by the stars
> d. all the above

Using radio telescopes that detect the 21-cm radiation emitted by atomic hydrogen, astronomers have discovered that most of the hydrogen in our galaxy is distributed in the galactic disk, along the spiral arms.

Visible light is scattered and absorbed by dust, while infrared and the 21-cm radio radiation travels freely through dust clouds.

The interstellar matter in the Milky Way accounts for about 10% of the mass of the all the stars in the galaxy.

Molecular Clouds and Star Formation

In chapter five, we learned that the solar system formed from a collapsed molecular cloud. In this chapter, we will extend the discussion to include other stars. Chapter five also described the main properties of molecular clouds.

The density of the molecular clouds is larger than the density of the diffuse interstellar matter. For this reason, astronomers say that molecular clouds are dense clouds of gas and dust. To see why this is so, compare the average density of a molecular cloud of about a million molecules per cubic centimeter (10^9 molecules/m^3), with the density of the IM of only 7 atoms/cm^3.

The molecular clouds are part of larger systems called **giant molecular clouds**. These large systems are not uniform and have the tendency to form small dense molecular cores. The tips of the pillars are examples of molecular cloud cores that might be collapsing to form stars.

Astronomers call these cores **evaporating gaseous globules** because the heat of young and hot stars is boiling off the surrounding gases.

Approximately 6,000 giant molecular clouds may populate the Milky Way and have masses between 100,000 to a million solar masses. The temperature of these systems is very low, ranging approximately 10–25 K.

Because the molecular clouds contain a large number of dense cold molecular cores, the stars usually form in **clusters**.

The following paragraphs contain a discussion of the formation of a star similar to the Sun. The molecular core giving birth to a Sun-like star has a small initial rotation speed of about 1 km/s. It has more than 1,000 solar masses, has a temperature of 10–50 K, and is about 1 parsec in diameter, and has a composition of 73% of H, 25% of He, and 2% of other elements.

Molecular Clouds and Hydrostatic Equilibrium

The different molecules constituting the molecular clouds' cores move in different directions with different speeds. The net effect of this motion causes the molecular core to slowly rotate around its axis. The internal motion of the molecules produces an outward pressure, attempting expand and destroy the molecular. This outward force is balanced by the inward gravitational pull.

Under these two forces, the molecular cores and the molecular clouds in general, are in hydrostatic equilibrium. See Figure 7-6. However, when the equilibrium is disrupted by a shockwave, the clouds' cores collapse under the gravitational pull and star formation begins.

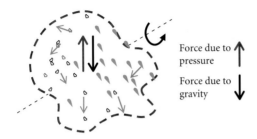

FIGURE 7-6. *Molecular clouds are in hydrostatic equilibrium the force due to gravity balances the force to pressure.*

In chapter five, we saw that the shock wave that triggers the collapse of a molecular core might be produced by one of the following:

1. when a nearby supernova explodes
2. when a nearby massive star is born
3. when a star passes close to the molecular cloud when a molecular cloud enters the arms of the Milky Way. (The arms of the Milky Way are regions of slowly traveling shock waves.)

From a Molecular Cloud Core to a Protostar

Only cold molecular cores collapse under the gravitational pull. As a core collapses, its temperature and its speed of rotation are gradually increased.

As matter collapses, it converts potential energy into kinetic energy. Therefore, the atoms gain speed and the temperature is increased. The higher the speed, the higher the temperature.

Recall that the average kinetic energy of the atoms in a gas gives the gas a specific temperature. Half of the initial potential energy is converted into heat, or kinetic energy, of the particles, the other half is radiated away in the infrared as heat.

As the core collapses and shrinks, it rotates faster as required by the conservation of angular momentum.

The collapsed core rotates faster and is hotter than the precollapsed molecular core.

Initially the collapse of the molecular cloud core is a "free fall," but as the temperature, the turbulence, the pressure, and the magnetic field increase, the contraction slows down.

Q13. Emission nebulae glow red/pink because _____.

a. they contain iodine
b. they reflect more efficiently the red light than any other color
c. the UV radiation emitted by the stars ionizes the hydrogen atoms
d. when free electrons recombine with the excited hydrogen atoms they emit a red–pink glow
e. c and d

Q14. Molecular clouds can be mapped using _____ that detect the radiation emitted by the CO + 1 ions.

a. Optical telescopes
b. Binoculars
c. Radio telescopes
d. X-ray telescopes

Q15. In the molecular clouds, most of the hydrogen is in the form of _____.

a. neutral atoms
b. negative ions
c. molecules
d. a and b

Q16. The _____ of the molecular clouds' cores that collapse under their gravitational pull to give birth to stars, is close to _____.

a. average size; 2 AU
b. average density; 1012 g/cm^3
c. mass; 0.2 solar masses
d. average temperature; 10 K

Most of the falling matter forms a hot central core that will ultimately give rise to a star, but some of it also accretes into a rotating disk around the center. This disk is where planets might form, as was the case of our solar system. See Figures 7-7 and 7-8.

Early in the process, the collapsed system remains cold and emits energy to space mainly in the infrared. However, as its internal temperature and pressure increases, the system becomes dense enough to prevent radiation to escape. The system becomes opaque to the radiation, which means that it is no longer transparent to the radiation. This is a critical phase in the collapsed system. When the system becomes opaque, it no longer emits energy from its bulk, rather energy is emitted only from its surface or **proto-photosphere**. This early stage in the star formation is called a **protostar**.

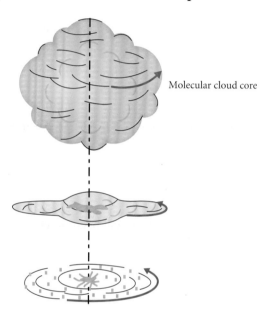

FIGURE 7-7. *A slowly rotating molecular cloud collapses unto a disk.*

FIGURE 7-8. *A young protostar in the Orion Nebula observed by the Hubble Space Telescope, surrounded by a protodisk.*

Credit: NASA Hubble Space Telescope Collection

From Protostar to Main Sequence

As the protostar contracts, it converts gravitational (potential) energy into heat. When the protostar becomes opaque, its internal pressure and temperature increase at a faster rate than on its surface.

Q17. Normally the molecular clouds are in hydrostatic equilibrium, but a _____ might disrupt the equilibrium and the cloud will collapse under its gravitational pull.

a. black hole

b. cloud shake

c. shock wave

d. heat wave

Q18. The protostars are still contracting and growing hotter because they convert gravitational (potential) energy into heat.

a. True

b. False

When the surface temperature is roughly 4,500 K and the central temperature is approximately 5,000,000 K, the system is called a protostar. Protostars have low surface temperatures. Therefore, they emit mainly in the infrared.

Because protostars are contracting, they are not in hydrostatic equilibrium. Thus, the outward force of pressure is less than the inward gravitational force.

The initial evolution of the protostars is hidden by a shroud of gas and dust. If we could see through this shroud of dust, we would see a huge contracting protostar more luminous than the main sequence star of the same temperature glowing in the infrared.

Protostars are systems evolving toward main sequence. They are not yet main sequence because they are still contracting, they are not in hydrostatic equilibrium, and they are not producing energy by a nuclear reaction.

As the protostars evolve, many become unstable and their luminosity becomes erratic. This stage in star formation is called **T-Tauri**. The properties of the T-Tauri star will be explained in the next section.

After the T-Tauri phase, the contraction of the protostar continues, until its internal temperature and pressure are high enough to halt the contraction.

A protostar, of about 1 solar mass, will begin to create energy by a nuclear reaction when its core temperature is about 10 million K. However, this temperature is not high enough to produce the pressure needed to stop the protostar's contraction. Only when the core's temperature reaches 15 million K, the gas pressure is strong enough to balance the gravitational pull. At this point, the star stops contracting. It has finally reached hydrostatic equilibrium and has become a main sequence star.

In the HR diagram, the different set of points representing the change of luminosity and temperature of a protostar is called the **evolutionary track**. Figure 7-9 gives the evolutionary track of a protostar of 1 solar mass.

Figure 7-10 shows the evolutionary track of four pre-main sequence stars. The diagram clearly shows that the time a protostar takes to become a main sequence star depends on the mass of the star. For example a 9 solar mass protostar takes only 150,000 years, while a 1 solar mass protostar takes only 10 million years.

Whatever the mass of the collapsing molecular cloud core, the endpoint of the prestellar evolutionary track is the main sequence band.

The main sequence band predicted by theory is called the **zero-age main sequence**.

A protostar with less than 0.8 solar masses will not be able to produce the 10 million degrees needed to initiate the fusion of

Q19. **The average number of particles per cubic centimeter in a molecular cloud is _____ the number of particles per cubic centimeter in Earth's air.**

a. comparable to

b. much less than

c. much more than

Q20. **The best place to look for stars in formation is _____.**

a. in the halo of the Milky Way

b. in a region rich in red giants

c. in a region that contains young and hot stars

d. inside molecular clouds

e. c and d

Q21. **Which of the following describes an interstellar molecular cloud?**

a. It can have many molecular cloud cores.

b. It has a temperature between 10 and 50 K.

c. It can give rise to a cluster of stars.

d. All the above

Q22. **If you were to look for a protostar, you would use an _____ telescope.**

a. optical

b. ultraviolet

c. infrared

d. X-ray

Q23. A protostar becomes a main sequence star when _____ .

a. it begins to make energy by a nuclear reaction
b. the surface temperature is of the order of 2,000 K
c. it stops contracting and reaches hydrostatic equilibrium
d. the core temperature is about 15 million K
e. c and d

Q24. Which of the following protostars will first begin the production energy by a nuclear reaction?

a. a 10 solar mass
b. a 20 solar mass
c. a 2 solar mass
d. a 1 solar mass
e. a 0.5 solar mass

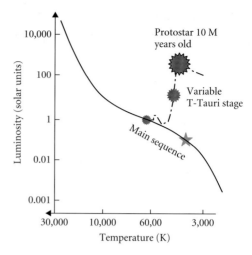

FIGURE 7-9. *Evolutionary track of a 1 solar mass protostar.*

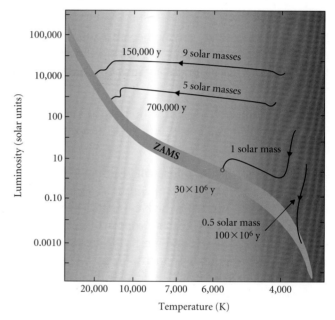

FIGURE 7-10. *The time a protostar takes to become a main sequence depends on the mass of the star.*

hydrogen in the core. Therefore, such protostars will never become a star rather the object formed is a brown dwarf.

Brown dwarfs are failed stars and are small, faint, cool (and growing ever colder), and difficult to detect. They are called **failed stars** because the mass of the collapsed core is insufficient to produce the pressure and temperature to generate energy by nuclear fusion.

Three important characteristics of main sequence stars:

1. They are in hydrostatic equilibrium. This means that the rate at which the star produces energy in the core equals the rate at which the star radiates energy from its surface or photosphere.

2. They produce energy at a constant temperature by fusing hydrogen into helium in the core. The core temperature

of stars (like the Sun) is between 15 to 17 million K. More massive stars have a higher core temperature of up to 27 million K.

3. The luminosity, surface temperature, and size remain fairly constant.

Now we turn our attention to T-Tauri stars.

CHARACTERISTICS OF T-TAURI STARS

T-Tauri stars are premain sequence with the following characteristics:

1. They are young objects still contracting and converting gravitational energy into heat.

2. They are located near or inside molecular clouds, and most of them are in binary systems with less than 3 solar masses.

3. They have lithium in their atmospheres. The presence of lithium is indicative of their extreme youth (less than 10 million years old) as lithium is rapidly destroyed in stellar interiors. If they were old stars, their spectra would not contain lithium lines.

4. They are still accreting extensive disks from the nebulae in which planets might form.

5. They are intermediate objects, lying in the evolutionary track between protostars and main sequence. They are rather cool, of spectral type G–M, but bigger and more luminous than the main sequence stars of the same spectral class. See Figure 7-9.

6. They are variable stars in that their luminosity fluctuates erratically. The change in brightness may be due to instabilities in the disk and violent activity in the stellar atmosphere. Their light curve does not show any periodic behavior as shown in the light curve of Figure 7-11. (A light curve is a plot of the luminosity or brightness of a star as a function of time.)

7. They may have huge dark regions similar to the sunspot, and a strong solar wind that clears the protostar from the gas and dust that did not accrete to the protostar. Protosars are visible only when they eliminate the shroud of dust that surrounds them.

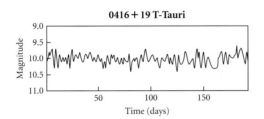

FIGURE 7-11. *Erratic change in the luminosity of a T-Tauri star.*

Use Figure 7-20 to answer questions 25 through 27.

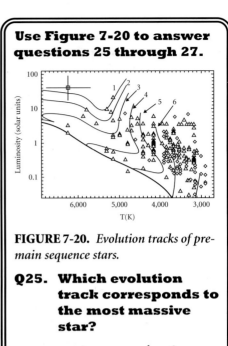

FIGURE 7-20. *Evolution tracks of pre-main sequence stars.*

Q25. Which evolution track corresponds to the most massive star?

a. 1 b. 2

c. 3 d. 4

e. 6

Q26. Which evolution track corresponds to the least massive star?

a. 1 b. 2

c. 3 d. 4

e. 6

Q27. Which evolution track corresponds to a star like the Sun?

a. 1 b. 2

c. 3 d. 4

e. 6

Q28. Herbig-Haro objects are associated with _____.

a. main sequence stars

b. brown dwarf objects

c. young protostars

d. formation of planets

Q29. The appearance of the bipolar flow and star wind in protostars _____.

a. causes the luminosity of the protostar to fluctuate erratically

b. triggers the fusion of hydrogen into helium

c. indicates that the protostar has become a main sequence star

d. indicates that the protostar is in hydrostatic equilibrium

Q30. T-Tauri stars are _____.

a. premain sequence stars

b. still contracting

c. a type of variable star

d. all the above

e. a and b

Q31. T-Tauri stars are _____ than the main sequence star of the same temperature.

a. older

b. smaller

c. more luminous

d. younger

e. c and d

Q32. T-Tauri stars _____.

a. are more luminous than main sequence star of the same spectral class

b. have a core temperature 15 million K and are fusing hydrogen

c. have constant luminosity

d. are post main sequence stars

e. a and c

Some active T-Tauri stars have two jets of gas going out perpendicularly to their disks. The jets originate in their atmospheres and are known as **bipolar flow** or **Herbig-Haro filaments**. See Figure 7-12.

As the bipolar jets move away from the protostars, they collide with the surrounding molecular cloud producing diffuse nebulae called **Herbig-Haro objects**. See Figure 7-12.

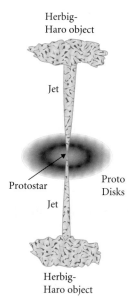

FIGURE 7-12. *Bipolar flow from a protostar and Herbig-Haro objects.*

Most of the Herbig-Haro objects are visible in the infrared. Only a few of them are observed in the visible part of the electromagnetic spectrum.

Notice that protostars giving rise to jets of gas are hidden by dust.

The appearance of the bipolar flow and star wind in protostars usually triggers the erratic changes in the luminosity of T-Tauri stars. The bipolar flow and the Herbig-Haro objects last only a few million years, and they disappear as the protostars continue their inexorable journey to the main sequence.

The first prototype of these variable stars was discovered in the constellation Taurus, and thus it was called **T-Tauri**. This T-Tauri consist of a binary system.

STARS SPEND MOST OF THEIR LIVES AS MAIN SEQUENCE

The period of time stars exist as main sequence, burning hydrogen into helium, depends on the mass of the star. Independent of their mass, stars spend about 90% of their lives as main sequence.

In general, low and medium-mass stars stay longer as main sequence than high-mass stars. High-mass stars consume fuel very quickly in order to produce the large amounts of energy required to keep up the large internal pressure. That large internal pressure supports their enormous weight. Therefore, these stars use their fuel (mass at the core) at a faster rate than the less massive stars. The stellar life-time, or time the stars stay as main sequence, is given by

$$\text{Stellar lifetime} = \frac{\text{Available fuel}}{\text{Rate of fuel consumption}}$$

The fuel of the stars is proportional to their mass, and the rate of consumption is proportional to the luminosity **L**. (Recall that $L = M^{3.5}$.) If the mass is in solar units, the stellar lifetime will be in solar lifetimes or 10 billion years. Doing the math, we get

$$\text{Stellar lifetime} = \frac{10}{m(\text{solar mass})^{2.5}} \times 10^9 \text{ years} \quad (7\text{-}1)$$

Giant and white dwarfs do not obey relation **7-1**.

Example

What is the stellar life span of a 10 solar mass O type main sequence star?

Solution

$$\text{Stellar lifetime} = \frac{10}{10^{2.5}} \times 10^9 \text{ years}$$

$$= \frac{10}{316.2} \times 10^9 \text{ years}$$

$$= 0.0316 \times 10^9 \text{ years}$$

$$= 31.6 \times 10^6 \text{ years}$$

Using equation 7-1, you will discover that a 2 solar mass main sequence star has a stellar life of roughly 1.36 billion years.

The results of these two examples and problems 36, 37, and 38 indicate that high-mass stars live only a few million years, medium-mass stars a few billion years, and low-mass stars many billion years.

The low-mass main sequence red dwarf stars have very long stellar lifetimes. For example, a red dwarf of 0.8 solar masses would remain as main sequence about 17 billion years. (See problem 38.) This is a long time. The universe is only 13.7 billion years old. So not a single red dwarf with less mass than 0.8 solar masses has ever left main sequence. Therefore, red dwarfs are the most abundant main sequence stars. (White dwarfs are another type of common star).

The O- and B-type massive stars are the least abundant because they go through their entire lives in a few million years. See Table 7-1.

Medium-mass stars (like our Sun) shine for a few billion years before they die. These stars are more abundant than the O and B high-mass

Q33. Why do astronomers rely heavily on infrared observations to study protostars?

a. Because the protostars are very luminous

b. Because the protostars surface temperature is rather low

c. Because the protostars are very old

d. Because the protostars have little hydrogen

Q34. One indication that protostars are young is that their spectra reveal the presence of _____.

a. lithium in their atmospheres

b. hydrogen

d. carbon

c. oxygen

e. all the above

Q35. How are Herbig-Haro objects related to star formation?

a. They are not related.

b. They are observed in regions rich in young stars.

c. They are a kind of main sequence stars.

d. They are always found near or inside molecular clouds where stars are forming.

e. b and d

Q36. The stellar lifetime of a 4 solar mass (medium-mass) star is ____ years.

a. 31,206
b. 10 million
c. 0.3 billion
d. 2 billion
e. 312 billion

Q37. The stellar lifetime of a 1.0 solar mass (medium-mass) star is ____ years.

a. 10 billion
b. 10 million
c. 5.6 million
d. 5.6 billion
e. 56 billion

Q38. The stellar lifetime of a 0.8 solar mass (low-mass) stars is ____ years.

a. 31,206
b. 10 million
c. 5.6 million
d. 5.6 billion
e. 17 billion

Q39. The Sun and the stars produce energy in their _____.

a. surfaces or photospheres
b. convection zone
c. atmosphere
d. core or near it

TABLE 7-1. *Lifetime of some main sequence stars.*

SPECIAL TYPE (TEMPERATURE)	SOLAR MASS	LIFETIME (YEARS)
O (35,000 K)	25	3.2×10^6
B (30,000 K)	15	1×10^7
A (11,000 K)	3	6×10^8
F (7,000 K)	1.5	4×10^9
G (580 K)	1	10×10^9
K (5,000 K)	0.8	17×10^9
M (3,500 K)	0.5	36×10^9

stars, but less abundant than the red dwarfs. See Table 7-1. Finally, when we look at the sky at night with the naked eye, most of the stars that are visible are O and B main sequence, giant, and super giant stars. Medium and low-mass stars are very difficult to see without a telescope.

NUCLEAR ENERGY GENERATION IN MAIN SEQUENCE STARS

Main sequence stars produce energy by means of a nuclear reaction, which converts four hydrogen ions into helium ions in the core, along with the liberation of energy. The reaction is written as:

$$4\,^1H^+ \longrightarrow {}^4He^{++} + energy$$

[Note: *The superscript indicates the number of protons plus neutrons. Helium has two protons and two neutrons. The exponent indicates that the atoms are ionized.*]

The helium nucleus, or alpha particle, has two protons and two neutrons. Therefore, the reaction needs four hydrogen nuclei to produce one helium nucleus. Each reaction liberates a small amount of energy by converting mass into energy.

The reaction is known as the **proton–proton reaction** because the intervening hydrogen atoms are ionized and the reaction occurs between bare protons.

The energy appears in the form of gamma (γ) ray photons, positrons (e^+), and neutrinos (ν).

The photons (γ) are absorbed by the hydrogen atoms before they travel very far. The absorption is followed by an emission of energy.

The positrons (e^+) readily combine with electrons, producing more gamma (γ) rays.

The neutrinos are particles with the following properties: they carry energy, travel almost at the speed of light, have no detectable charge or mass, and rarely interact with matter. Because the neutrinos rarely interact with matter, they are very difficult to detect. Each nuclear reaction inside the Sun-like star produces three neutrinos. Therefore, there is a huge flux of neutrinos flowing away from the interior of the Sun and the stars.

The nuclear reaction that fuses hydrogen into helium needs high temperatures, a minimum of 10 million K because the hydrogen nuclei (protons) need to move fast (large kinetic energy) in order to overcome the repulsion barrier (electromagnetic repulsion) between the protons.

Why does the fusion of hydrogen liberate energy?

The fusion of hydrogen liberates energy because it converts mass into energy. The mass of the four protons involved in the reaction is a little larger than the mass of the resulting helium nucleus. This means that the reaction converts a small amount of mass into energy, as predicted by Einstein's famous formula, **E = mc²**.

The missing mass per reaction is only 0.048×10^{-27} kg.

However, for each second in the Sun and in the stars, there are more than 10^{38} atoms of hydrogen nuclei fused into helium nuclei.

In 1 s, a star like the Sun converts 5 million tons of mass into energy and radiates it into space! A Sun-like star releases 10^{26} J every second or equivalent 10^{26} W.

Stars more massive than the Sun have a core temperature between 18 and 27 million K and fuse hydrogen into helium at a faster rate. To do this, the stars mix the hydrogen nuclei with carbon, nitrogen, and oxygen. This reaction is known as **CNO cycle**. These elements appear intact in the final reaction—their mission is to speed up the reaction. They are called **catalytic elements**.

The proton–proton reaction is efficient at temperatures of less than 18 million K, and the CNO reaction is more effective for temperatures between 18 and 27 million K.

As hydrogen is fused into helium, the number of hydrogen nuclei in the core of a stars gets lower. This occurs while the number of helium atoms increases. The fewer hydrogen nuclei must move faster to maintain the same pressure. However, this increases the core temperature, causing the fusion rate to increase, which also increases the star's luminosity or brightness. As the main sequence stars get older, their brightness increases incrementally. This change is very slow. Moses and Jesus saw the same brightness in the Sun that we see now. Main sequence stars are depleting hydrogen in the core and also building a helium core.

The energy that the stars form is produced in the core. From the core, the energy has to be transported to the surface and from the surface to space. Let's look at the mechanism responsible for transporting the energy inside the stars.

Q40. The Sun and the stars produce energy _____.
a. by converting gravitational energy into heat
b. by means of a chemical reaction on its surface
c. by oxidation of carbon in the core
d. by fusing hydrogen into helium in or near the core
e. b and d

Q41. The outward pressure of hot gas in the stars is balanced by _____.
a. the inward gravitational force
b. increasing the star's diameter
c. cooling the photosphere
d. changing their luminosity
e. increasing the star wind

Q42. In the proton–proton cycle, the helium nuclei and the neutrinos have less mass than the original four hydrogen protons. What happens to the "lost" mass?
a. It is recycled back into hydrogen.
b. It is ejected into space.
c. It is converted to energy.
d. It is transformed into electrons.
e. Conservation of mass dictates that no mass can be lost.

Q43. Why is "convection" important in the stars?

a. It speeds the production of energy.

b. It determines how fast molecular clouds collapse.

c. It mixes the gases in the stars.

Q44. Which of the following does not describe a property of a main sequence star?

a. It fuses helium into carbon.

b. It is in hydrostatic equilibrium.

c. Its luminosity is fairly constant.

d. It fuses hydrogen into helium in the core.

Q45. The core temperature of a star like the Sun is _____ million K.

a. 1

b. 10

c. between 15 and 17

d. 22

Q46. The core temperature of a star several times the mass of the Sun is _____ million K.

a. 1

b. about 10

c. less than 15

d. between 17 and 27

Energy Transport Inside the Stars

In general, energy can be transported by conduction, radiation, and convection. The energy inside the stars is transported by radiation and convection. See Figure 7-13.

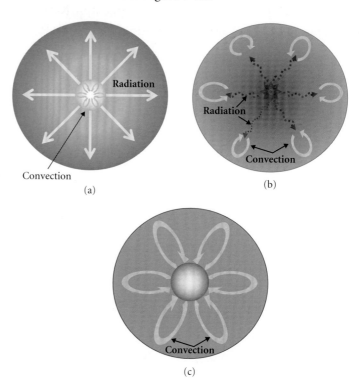

FIGURE 7-13. *Conduction of energy in the interior of the stars. (a) High-mass stars, (b) medium-mass stars, and (c) red dwarfs.*

By radiation, the energy is carried by photons. By convection, energy is carried by the movement of matter from regions of higher temperature to regions of lower temperature.

Medium-mass stars, like the Sun, with solar masses between 0.8 and 8, transport the energy from the core to the surface first, by radiation in the inner layers, followed by convection.

In high-mass stars, with more than 8 solar masses, the energy is transported first by convection and then by radiation due to the higher pressure and density of the star.

Red dwarf stars, which have solar masses between 0.4 and 0.08, transport energy only by convection. These stars are fully convective. Convection mixes the helium produced in the core with the hydrogen all over the body of the stars.

From the surface (photosphere) of all stars, the energy is transported to space by radiation.

When stars deplete the hydrogen in their core, their cores contract and increase in temperature. After this, they begin to fuse heavier elements than hydrogen. We will describe this new phase in the lives of the stars in our next chapter. The most important star for us is our Sun, so we are going to have a closer look to this star.

A Brief Look at Our Star: The Sun

The Sun is the closest star to Earth. The Sun is intensively studied by the **Global Oscillation Network Group**, located in different places on Earth, with the two telescopes of the National Solar Observatory. One of these telescopes is located in New Mexico, and the other in Arizona, (http://www.nso.edu/). From space, the Sun is monitored by the Solar and Heliospheric Observatory (SOHO) and other spacecrafts.

SOHO is in orbit around the Sun, and from this position, it enjoys an uninterrupted view of our daylight star (http://sohowww.nascom.nasa.gov/).

The photosphere is the visible surface of the Sun. The photosphere is opaque, and we cannot see the interior of the Sun. Above the photosphere lies the Sun's atmosphere. The lower layers of the atmosphere form the chromosphere, and the top layers of the atmosphere form the corona. See Figure 7-14. The photosphere is dotted with a network of bubbles, or granules, produced when hot matter from the interior breaks onto the Sun's surface. Each granule lasts about 1 h and has a diameter of about 1,000 km.

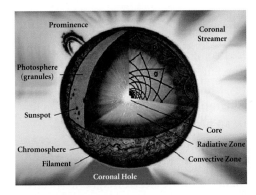

FIGURE 7-14. *Sun's structure.*

Credit: NASA Solar and Heliospheric Observatory Collection

This granulation is due to the convection operating below the Sun's photosphere. The Sun is an active star. The activity is linked to its differential rotation and to the presence of a large magnetic field on its surface. This magnetic field is due to the **dynamo effect**. The Sun's magnetic activity seems to be linked, among other phenomena, to sunspots, solar flares, prominences, and solar wind.

Sunspots

Sunspots are regions where the magnetic field is larger than on the rest of the photosphere. Each sunspot is larger than the Earth and appears black because it is cooler (at only about 4,500 K) than the surrounding material on the photosphere. See Figure 7-15.

The sunspots move on the surface of the Sun, as the Sun rotates on its axis. Galileo observed the rotation of the sunspots and concluded that the Sun was rotating on its axis. Further, they are not

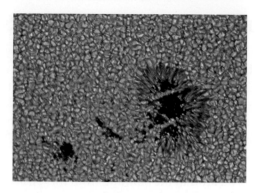

FIGURE 7-15. *The Sun's photosphere and sunspots.*

Credit: NASA Space Fly Center. http://solarscience.msfc.nasa.gov/feature1.shtml

permanent. Some sunspots last a few hours and others persist for months. The number of sunspots varies with a mysterious cycle of 11 years.

The spots usually occur in pairs because one spot is associated with the south magnetic pole and the other associated with the north magnetic pole.

Solar Flares

Solar flares are bright and powerful bursts of hot ionized gas that originates in the Sun's photosphere near the sunspots. Solar flares get their energy from the large magnetic field associated with the sunspots. The flares rise to their maximum in a few minutes and decay in about an hour.

The gases within the solar flares have temperatures of up to 5,000,000 K. Consequently, they radiate energy in virtually all the different wavelengths of the electromagnetic spectrum, from radio waves to gamma rays. They are great sources of UV and X-Ray radiation.

They contribute a large number of charged particles to the solar wind.

Prominences

The prominences, shown in Figure 7-16, are huge loops of gas that originate on the Sun's photosphere and extend high into the upper atmosphere to the Sun's corona. The prominences are not as violent as the solar flares.

Some prominences last only a few days, and the most stable persist for weeks and months. The prominences are also associated with sunspots. The prominences can be seen during a total solar eclipse using the appropriate filters on telescopes.

The stars are very far away, and we cannot see their surfaces or photospheres in detail. The only stars, other than the Sun, whose photospheres have been imaged by the Hubble Space Telescope, are Mira Ceti and Betelgeuse.

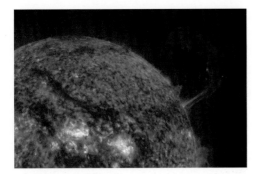

FIGURE 7-16. *Prominences.*

Credit: NASA Marshall Space Flight Center Collection

Solar Wind

The **solar wind** is a stream of charged particles (plasma) that escape from the upper layers of the Sun's atmosphere. The solar wind reaches to the outermost part of the solar system.

The solar wind and the Sun's radiation cause the tails of comets to always point away from the Sun.

Composition of the Sun

Analysis of the solar spectral lines shows that the Sun contains, by mass, about 71% of hydrogen, 27% of helium, and 2% of other elements such as oxygen, carbon, nitrogen, silicon, magnesium, neon, iron, sulfur, and others. The Sun's composition is very similar to other stars.

STAR CLUSTERS

The stars have the tendency to form into clusters. The Open cluster and the Globular cluster are two important clusters. The stars in a cluster were formed from the same molecular cloud at approximately the same time. Therefore, they are roughly the same age, the same composition, and the same distance from Earth. However, the stars in a cluster have different masses, different sizes, different luminosities and are at different stages in their evolution.

A young group of stars loosely bound by their mutual gravitational interaction is called an **open cluster**. They are always near the location where they were formed, which is near or inside molecular clouds, indicating that they are young systems.

A good example is the open cluster in the 30 Doradus, or Tarantula Nebula, located in the Large Magellanic cloud.

In our Milky Way galaxy, the open clusters are located in the galactic disk where the molecular clouds and HII regions are located. Over 1,100 open clusters are known in our Milky Way galaxy, but many more are yet to be discovered.

Q47. In the following list, what does the physical nature of solar granules most resemble?

a. a pot of bubbling fudge

b. ripples on a pond

c. small dark holes

d. none of the above

Q48. The only absorption lines present in the solar absorption spectrum are the lines caused by hydrogen.

a. True

b. False

Q49. The mass of the Sun has the following composition:

a. approximately 27% of hydrogen, 71% of helium, and 2% of other elements

b. approximately 71% of hydrogen, 27% of helium, and 2% of other elements

c. approximately 71% of oxygen, 27% of helium, and 2% of other elements

Q50. The presence of _____ in the Sun's photosphere gives evidence of solar convection.

a. flares

b. granules

c. prominences

d. b and c

Q51. A typical solar flare lasts _____.

a. a month

b. a few days

c. about an hour

d. 20 s

Q52. Figure 7-21 represents five star clusters of different ages. Classified them from youngest to oldest.

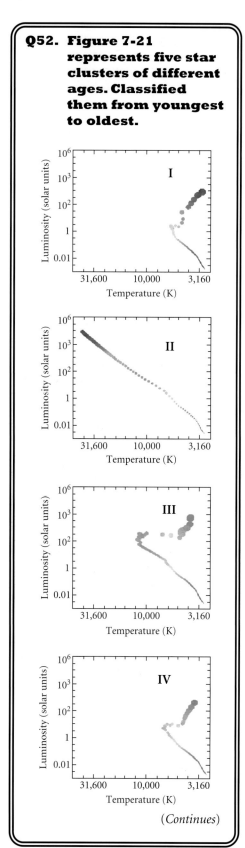

(*Continues*)

The mutual gravitational force between the individual stars in open clusters is weak. Therefore, they do not have a regular shape, and they disperse in several million years. Only a few of them reach ages of a few billion years.

Long ago, our Sun was a member of an open cluster, but it left the nest where it was born. Most of the stars in the Hyades open cluster in the constellation Taurus are main sequence stars, but they also contain evolved old red giants. Open clusters contain about 1,000 stars and are less than 10 pc in diameter.

Globular Clusters

In contrast to open clusters, the globular clusters are ancient, approximately 10 billion years old. The stars in these clusters have a strong gravitational interaction and, therefore, have regular spherical spatial distribution. Globular clusters contain anywhere from a few thousand to a million stars, which are distributed within a spherical region of 20–100 pc in diameter.

They are located in the halo and bulge of the galaxy, far away from the arms and disk. About 150 globular clusters have been identified in the Milky Way. Most of the stars in globular clusters are low-mass main sequence, red dwarfs, and red giants. Most of the yellow variable RR-Lyrae are located in these types of clusters. The globular clusters lack upper massive main sequence stars.

The HR diagram of a star cluster gives important information about the cluster. We know that the stars in a globular cluster were formed at about the same time but have different masses and different luminosities. We also know that high-mass stars evolved faster than medium- and low-mass stars (See Table 7-1). The HR diagram is a snapshot of the evolution of the stars. A very young cluster has several protostars approaching the main sequence, and most of the stars within the cluster are main sequence, as shown in Figure 7-17a. In an older cluster, the majority of the stars are main sequence with a few protostars and a few red giants. See Figure 7-17b. In a very old cluster, the majority of stars have evolved and become red giants and white dwarfs and will not have any protostars. See Figure 7-17c. The point in the HR diagram where the stars leave the main sequence band is called the **turnoff point**.

The **Pleiades** (Messier object 45) is a young open cluster of only 150 million years. Its HR diagram indicates that most of the stars are still main sequence, below the turnoff point, and a few, the most massive, have evolved from main sequence and become giants, above the turnoff point. See Figure 7-18. The Hyades and the Beehive or Praesepe (M44) are examples of other well-known open clusters.

The 47-Tucanae (47-Tuc) is an example of globular cluster, located in the southern hemisphere at a distance of 13,000 ly. This cluster is also known as **NGC 104** and is approximately 10 billion years old and many stars have left the main sequence becoming red giants. Figure 7-19 shows the HR.

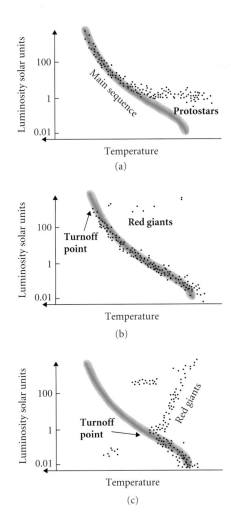

FIGURE 7-17. *Theoretical evolution of a star cluster.*

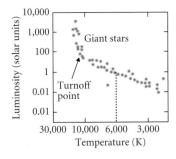

FIGURE 7-18. *HR diagram of the open cluster, the Pleiades (M45).*

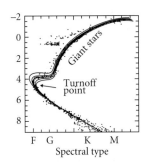

FIGURE 7-19. *HR diagram of the 47-Tuc globular cluster.*

(Continued)

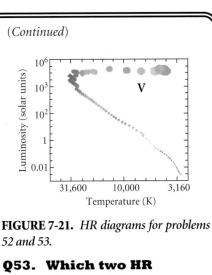

FIGURE 7-21. *HR diagrams for problems 52 and 53.*

Q53. Which two HR diagrams in Figure 7-21 represent open cluster? Why?

Q54. Which of the following is not true?

a. Most of the stars in open clusters are very young.

b. The globular clusters are located mainly in the disk and arms of the Milky Way.

c. The open clusters are located mainly in the disk and arms of the Milky Way.

d. The majority of the stars form in clusters.

e. c and d

Chapter 7—The Lives of the Stars: From Birth to Main Sequence Stars

Q55. Which of the following is a characteristic of open clusters?

a. Most of the stars are red giants.
b. Very young clusters have hot O and B luminous main sequence stars.
c. Some clusters have young and old red giants stars.
d. All the stars have the same mass.
e. b and c

Q56. This cluster has several young luminous stars, is about 180 million years old, and has a few red giants. This describes _____.

a. an HII region
b. a molecular cloud
c. an open cluster
d. a globular cluster

Q57. The lower the turnoff point in the H-R diagram of a star cluster, the _____.

a. younger the cluster is
b. older the cluster is

Answers

1. c
2. a
3. d
4. b
5. c
6. b
7. d
8. ultraviolet; hydrogen
9. d
10. c
11. e
12. a
13. d
14. c
15. c
16. d
17. c
18. a
19. b
20. e
21. d
22. c
23. e
24. b
25. a
26. e
27. d, The main sequence star at the end of the track has a temperature of 5,800 K and a luminosity of 1 solar unit.
28. c
29. a
30. d
31. d
32. a
33. b
34. a
35. e
36. c
37. a
38. e
39. d
40. d
41. b
42. c
43. c
44. a
45. c
46. d
47. a
48. b
49. b
50. b
51. c
52. II, V, III, IV, and I
53. III and IV. III is the youngest (all stars are main sequence). In IV, only a few stars have evolved from main sequence.
54. b
55. e
56. c
57. b

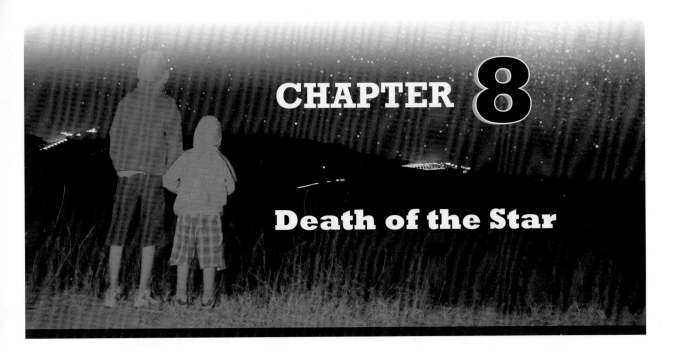

CHAPTER 8

Death of the Star

Because the mass of the stars determine the rate of evolution of the stars, we divided the main sequence stars into three different groups according to their masses as follows: low mass, medium mass, and high mass stars. See Figure 8-1.

LOW MASS STARS OR RED DWARFS

Low mass stars, also called **red dwarfs**, have masses between 0.08 and 0.4 solar mass. These are the most abundant kinds of main sequence stars. Some astronomers believe that about 80% of the stars in the galaxy and perhaps in the universe are red dwarfs. Because they have low temperatures (spectral type K and M) and low luminosities, they are invisible to the naked eye. Most of the red dwarfs discovered are within the first dozen or so parsecs of the solar system.

Red dwarfs are nearly fully convective. Convection is important in the stars because it mixes the helium produced by fusion in the core with the hydrogen gas over the bulk of the star. See Figure 8-2.

Since red dwarfs are almost totally convective, the helium that is made in the core is mixed with the hydrogen in the body of the star, which causes fresh hydrogen to be continually transported to the hydrogen burning core.

At the end of the hydrogen cycle, which lasts more than 14 billion years, the stars gradually cool down to become dark dwarfs.

Many red dwarfs are variable stars in that they irregularly emit intense flares with wavelengths that range from radio, visible, ultra-violet, and X-ray. The flares last from a few minutes to a few hours.

In chapter 7, we saw that our Sun also emits flares, but their energy is small compared with the Sun's luminosity. This is not the case for red dwarfs.

The amount of X-rays emitted by flare stars is thousand times more than that emitted by our Sun in a flare.

Some flare stars seem to have planets. Can any of these planets harbor Earth-type life?

Red dwarfs are cool, so for a planet to be hot enough to offer conditions similar to those on Earth, it has to be close to the dwarf. But being close to a flare star that emits X-ray radiation minimizes the chances for life. Furthermore, red dwarfs might be too red, and therefore, might not have the enough of the proper light to generate plant photosynthesis.

Q1. The most common main sequence stars are the _____.

a. red giants
b. stars like the Sun
c. super giant stars
d. red dwarfs

Q2. Convection is important in the stars because it

a. slows down the production of energy
b. speeds up the production of energy
c. mixes the gases in the stars
d. determines the rate of nuclear fission

Q3. Red dwarf stars _____.

a. will never fuse helium
b. fuse hydrogen in the core at a faster rate than more massive stars
c. make the largest helium core ever observed
d. give origin to the largest red giants in the galaxy

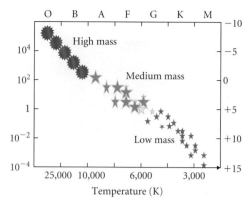

FIGURE 8-1. *Main sequence stars can be grouped into three different types according to their mass.*

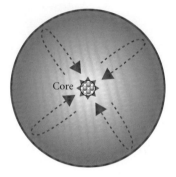

FIGURE 8-2. *Red dwarfs are totally convective. The helium they make in the core is mixed over the bulk of the stars.*

Proxima Centauri, the closest star to our solar system, is a flare star. The most well-known flare star is the UV Ceti, discovered in 1948. All red dwarfs that emit flares are known as **UV Ceti variables**.

Celestial objects with less than 0.08 solar masses are not stars because they are not able to fuse hydrogen. They are called **brown dwarfs**.

Brown dwarfs are cool objects that are gradually getting cooler and will, over long period of time, become black dwarfs.

Medium or Intermediate Mass Stars

Main sequence medium mass stars have masses between 0.4 and 8 solar masses. These stars are partially convective. Therefore, as they fuse hydrogen into helium, they build a helium core. See Figure 8-3.

As an example of the evolution of a medium mass star, we will describe the evolution of a Sun-like star.

This star spends about 10 billion years fusing hydrogen into helium, which in essence, is building an inert helium core.

At the end of the hydrogen cycle, the temperature of the newly formed helium core drops dramatically. When this occurs, the star

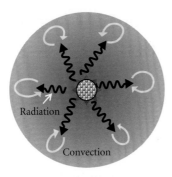

FIGURE 8-3. *Medium mass stars are partially convective.*

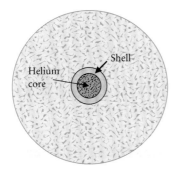

FIGURE 8-4. *When a star runs out of hydrogen in the core, it begins to fuse hydrogen in the shell and the star becomes a red giant.*

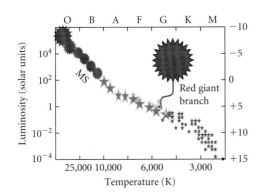

FIGURE 8-5. *Stars in the red giant branch are fusing hydrogen in the shell and its core is contracting and getting hotter.*

loses hydrostatic equilibrium and the helium core begins to contract. This increases its temperature and pressure. As the temperature in the core increases, the hydrogen gas around the inert helium core, or shell, becomes hot enough to fuse into helium. See Figure 8-4. Some of the heat produced through this process flows to the surface of the star, causing its atmosphere to expand and cool down. At this point, the star leaves the main sequence band and enters the red giant branch, becoming a red giant for the first time. See Figures 8-4 and 8-5.

The change from normal main sequence to a red giant is gradual and takes about 1 billion years.

Chapter 8—*Death of the Star*

> **Q4. Red giants ___.**
>
>   a. are high mass main sequence star
>   b. look red because they have cool atmospheres
>   c. are very luminous because they are very hot
>   d. b and c
>
> **Q5. Red giants are _____.**
>
>   a. the largest proto stars ever detected
>   b. postmain sequence stars
>   c. very luminous because they are large
>   d. b and c
>
> **Q6. Some red giants _____.**
>
>   a. are premain sequence stars
>   b. fuse hydrogen in their cores
>   c. use hydrogen in their shells
>   d. have a core temperature of 5 million K
>   e. b and c
>
> **Q7. What happens when a star can no longer fuse hydrogen to helium in its core?**
>
>   a. The core expands and cools off
>   b. The core shrinks and heats up
>   c. The core expands and heats up
>   d. Helium fusion immediately begins

Main Properties of Red Giants

When medium mass stars become main sequence for the first time, they are in the process of fusing hydrogen in their shell, their cores are shrinking and their temperatures are increasing. The envelopes of these stars are increasing during this process, and they are losing a large amount of mass in the form of a persistent star wind. Some loose between 20 and 30% of their original mass. (Stars much larger than the Sun can loose mass equal to several solar masses).

Some red giants have radii from 10 to 100 times the radius of the Sun, and surface temperatures of approximately 4,500 K or less. When the Sun becomes a red giant, it will probably expand to the orbit of Mercury and will occupy most of the sky during the day. The heat reaching the Earth will destroy life, the water of the oceans will evaporate, and everything will burn. When would this happen?

Even though red giants have low surface temperature, they are hundreds of times more luminous than the Sun because they are huge.

The red giant phase is short, only about 10% of the total star's lifetime.

Red giants are very old and evolve as postmain sequence stars closer to the moment of death than to the moment of birth.

Helium Cycle

The red giants, in the red giant branch shown in Figure 8-5, are fusing hydrogen in the shell, and their cores are contracting and increasing their temperature and pressure.

The temperature in the core is so high that all the atoms are ionized. The free electrons, now liberated by the atoms, are squeezed under the high core pressure. However, there is a moment in which the electrons would not get any closer. We say that the electrons become degenerate and the corresponding pressure is the degenerate pressure.

The properties of degenerate electronic gases are the subject of a branch of physics called **quantum mechanics**.

A gas of electrons, in which the free electrons are as close to each other as the laws of physics allow, is a degenerate gas in which the electrons become degenerate.

The refusal of degenerate electrons to be packed any closer provides the force that balances the gravitational pull.

The core of red giants are in hydrostatic equilibrium because the degenerate pressure balances the gravitational pull. This means that once the core of a red giant becomes degenerate, it does not contract any more although its temperature keeps increasing.

In normal gases, when the pressure increases, the temperature also increases, and vice versa. However, degenerate gases are not normal gases. Degenerate gases obey different rules than normal gases. For example, they can increase their temperature without increasing their pressure.

Because the core of red giants is degenerate, its temperature increases without increasing its internal pressure. When the core's temperature reaches 100 million K, the fusion of helium suddenly and explosively begins over the entire degenerate helium core. This is known as the **helium flash**.

The helium flash lasts only a few hours, and soon after the explosion, the pressure-temperature thermometer controls the core helium burning. After the helium flash, the star continues to fuse helium at a steady rate at a temperature well above 100 million K.

The energy liberated in the helium flash increases the gas pressure in the core, causing it to expand. The expanding core pushes the hydrogen burning-shell outward. A consequence of the expansion is that heat leaks to the surface of the red giant, which increases its temperature. Due to a mechanism that is not completely understood, even though the surface temperature increases, the volume of the giant decreases which lowers its luminosity. The giant then enters the horizontal branch and becomes a yellow giant, as shown in Figure 8-6.

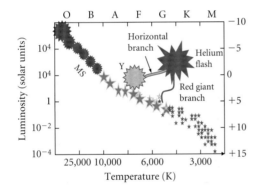

FIGURE 8-6. *Yellow giants are fusing helium in the core and hydrogen in the shell.*

Yellow giants are fusing helium into carbon at a temperature well above 100 million K. The fusion of helium produces carbon with the liberation of energy. Helium fusion is known as a **triple alpha process** because the reaction needs three helium nuclei or alpha particles to produce one carbon nucleus. The final reaction is as follows:

$$3\ [^4\text{He nuclei}] \rightarrow {}^{12}\text{C} + \text{energy}.$$

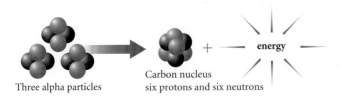

Q8. The helium flash _____.

a. is a sudden and violent ignition of helium in red giants

b. requires a temperature of 15 million K

c. happens on the surface of the red giants

d. triggers the silicon cycle in the red giants

Q9. The core of red giant star is degenerate. This means that the _____.

a. protons are recombining with the electrons

b. protons are as close to each other as possible

c. neutrons are as close to each other as possible

d. electrons are as close to each other as possible

e. protons are reacting with the neutrons

Q10. Some red giants _____.

a. are as big as Jupiter

b. are main sequence star

c. have surface temperatures of about 4,000 K

d. are fusing hydrogen in the core

Chapter 8—*Death of the Star*

Q11. At the end of the helium cycle, stars have built a _____.

a. rich helium atmosphere
b. hydrogen core
c. nickel oxygen core
d. carbon oxygen core

Q12. A helium nucleus or alpha particle has

a. two electron and two protons
b. two protons and two neutrons
c. four neutrons
d. two electrons and two neutrons

Q13. The fusion of helium involves three helium nuclei because the carbon nucleus produced in the final reaction has ____.

a. larger mass than the three alpha particles
b. six protons and six neutrons
c. less affinity than the alpha particles
d. eight protons and four neutrons

Q14. What happens in a Sun-like star when core temperature rises enough for helium fusion to begin?

a. carbon fusion begins
b. hydrogen fusion suddenly stops
c. helium fusion rises very sharply
d. the star blows away its core

---

helium fusion liberates energy because the mass of the three helium nuclei or alpha particles is slightly larger than the mass of the carbon nucleus. Thus, the reaction converts mass into energy.

This reaction also produces oxygen. Therefore, at the end of the helium cycle the star has built an inert carbon-oxygen core.

After the helium flash, the star makes energy in the helium-fusion core and in the hydrogen-fusion shell.

The helium cycle is short lived (approximately 100,000 years) for two reasons:

1. The hydrogen fusion reduces the number of atoms by one-fourth (Four hydrogen nuclei are needed to produce one alpha particle or helium nucleus). The helium cycle has one-fourth less nuclei to fuse than the nuclei available for the hydrogen cycle.

2. The rate of helium fusion goes at a pace faster than hydrogen fusion because the temperature required to fuse helium is about eight times higher (more than 100 million K for helium compared with 15 to 17 million K for hydrogen) than the hydrogen fusion temperature.

Many yellow giants, originating from medium mass stars like the Sun, are unstable and their luminosity begins to change rhythmically. These variable red giants are called **RR-Lyrae** variable stars.

We have already encountered another type of variable stars, the T-Tauri stars.

Before we continue the description of the evolution of the post main sequence stars, let's say a few words about the RR-Lyrae variable stars.

The RR-Lyrae are yellow giants whose luminosity increases and decreases rhythmically with a period between 12 and 24 h. All RR-Lyrae are pulsating yellow giants located in the horizontal yellow branch, after the helium flash.

All RR-Lyrae originate when main sequence stars, similar to the Sun, evolve and become yellow variables. Therefore, it is not surprising that all of them have about the same average absolute magnitude and similar period of variability in their luminosity. Figure 8-7 shows the light curve of a RR-Lyrae.

The apparent magnitude of each RR-Lyrae is different, because it depends on distance.

Since all RR-Lyrae have the same average absolute magnitude (about 0.7), measuring the apparent magnitude (the distance to these objects) can be found using equation 6-1.

RR-Lyrae variable stars are evolved old postmain sequence stars located mainly in globular clusters, which in turn are in the galactic halo of the Milky Way.

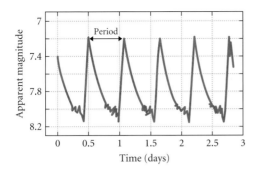

FIGURE 8-7. *The light curve of an RR-Lyrae variable star. Apparent brightness vs time.*

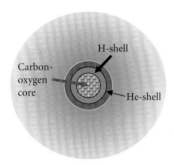

FIGURE 8-8. *Yellow giant, produced by an evolved on solar mass star.*

RR-Lyrae played an important role in the determination of the dimensions of the Milky Way as we will see when we study our galaxy.

RR-Lyrae variables are named after the prototype star RR variable found in the constellation Lyra. In the following section, we will encounter another type of yellow variable, mainly the variable yellow Cepheids. Now we will return to the evolution of the stars.

We left our Sun-like star in the horizontal branch fusing helium in the core and hydrogen in the shell.

At the end of the helium cycle, the star has built an inert carbon-oxygen core. As the helium is depleted, the core contracts which increases its temperature. The star then begins to fuse helium around the inert carbon-oxygen core, in the inner shell below the burning hydrogen shell.

In summary, the star at this point has a contracting carbon-oxygen core that is getting hotter and is fusing helium in the inner shell and hydrogen in the outer shell. See Figure 8-8.

As the carbon-oxygen core contracts, the temperature increases but it never reaches the 600 million K needed to begin the carbon cycle. Since the star never begins the carbon cycle, the envelope expands and cools, and the star enters the asymptotic branch becoming a red giant for the second time. This time, the red giant

Q15. What is the period of the RR-Lyrae whose light curve is given in Figure 8-7? _____.

a. 6 h
b. 0.6 days
c. 12 h
d. 1 day

Q16. The approximate change in magnitude of the RR-Lyrae whose light curve is given in Figure 8-7 is:

a. 1.4
b. 2
c. 1
d. 4

Q17. How do we determine the distance to RR-Lyrae variable stars?

Q18. A few RR-Lyrae variable stars are main sequence stars.

a. True
b. False

Q19. All RR-Lyrae variable stars have similar luminosities because they originate from

a. high mass stars
b. low mass stars
c. white dwarfs
d. stars with similar mass
e. brown dwarfs

> **Q20. When a star of 2 solar masses uses up all of the core's hydrogen**
>
> a. it probably begins to fuse helium in the core
>
> b. it probably begins to fuse hydrogen in the shell
>
> c. it probably collapses for lack of fuel
>
> d. the hydrogen from the envelope fills the empty core and reactivates the nuclear fusion
>
> e. gets cold and dies

is bigger and more luminous than before. At this point, the radius of the giant is about 500 times the radius of the Sun.

The contracting inert hot carbon-oxygen core liberates large amounts of heat that increase the temperature and activity of the two shells. This causes **shell flashes**, or **thermal pulses**, that separate the envelope (atmosphere) from the fusion shells. These flashes contribute to the expansion of the atmosphere of the red giant. The atmosphere drifts away in a gentle wind leaving behind a hot inert degenerate core of hot "carbon-oxygen" known as a **white dwarf**.

The ultraviolet radiation emitted by the hot carbon-oxygen core excites and ionizes the gases of the expanding nebula causing it to become visible as an emission planetary nebulae.

The planetary nebula gradually mingles with the interstellar gas, enriching it with carbon, oxygen, and in some cases, neon.

The stellar wind of red giants and planetary nebula carry away into space the dusty grains of carbon oxygen and other elements made in the core of the medium mass stars.

In approximately 5 billion years from now, our Sun will become a red giant and will end its life as a planetary nebula leaving behind a white dwarf.

Planetary nebulae are not related to planet formation. The astronomers who discovered them in the 18th century thought that they had discovered a giant planet. The name still persists.

The stars with masses larger than 2 but smaller than 8 solar masses can leave behind a white dwarf if they shed enough mass during the giant phases, during the thermal flashes, and when they form a planetary nebula.

This must be the case because the maximum mass a white dwarf can have is 1.4 solar masses. (See below for a discussion the white dwarf mass limit.)

The helium flash is characteristic of stars with masses between 0.4 and 2 solar masses. Stars with larger mass than 2 solar masses, gradually increase their core temperature and begin the helium cycle, without going through the helium flash.

Stars with mass between about 8–12 solar masses might get hot enough to fuse carbon into neon-oxygen, so when they lose their atmosphere in the form of a planetary nebula, they leave behind a neon-oxygen white dwarf. These types of white dwarfs are very rare.

In the H–R diagram of Figure 8-9, you can see the evolution (solid lines) of 1 and 5 solar mass stars as they leave the main sequence band. The 1 solar mass star will give rise to a white dwarf of about 0.6 solar mass and the other star to a white dwarf of 0.85 solar mass.

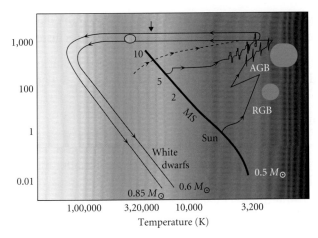

FIGURE 8-9. *Postmain sequence evolution of two medium solar mass stars.*

RGB means red giant branch, and AGB means asymptotic giant branch. The spikes in the diagram represent thermal pulses.

Before we study the evolution of high mass stars, let's turn our attention to white dwarfs.

Properties of White Dwarfs

White dwarfs are produced at the end of the evolution of medium mass stars so they are composed mostly of carbon and oxygen ions immersed in a sea of highly compressed degenerate electrons. Remember that degenerate electrons are electrons that are compressed in such a way that they are as close to each other as possible. The refusal of the electrons to get any closer produces the repulsion force that balances the gravitational pull in white dwarfs.

Detailed theoretical calculations made by Subramanyan Chandrasekhar show that electron degeneracy will not be able to support white dwarfs that exceed a mass of 1.4 solar masses. This maximum mass is known as the **Chandrasekhar mass limit**. The mass of white dwarfs goes between 0.5 and 1.4 solar mass.

The mass of the white dwarfs is packed in a volume as big as the Earth, which means that they have a large density. One cubic centimeter of white dwarf matter has a mass of 1 million grams (density of 10^6 g/cm³). Imagine the mass of the Sun packed in a volume as big as the Earth.

If we remember that white dwarfs are the degenerate core left by dying stars, it is easy to see why they initially have temperatures of more than 100,000 K. Since they do not make energy, they eventually cool down and become black dwarfs. In chapter 6, we saw that white dwarfs are not that luminous. Now we see why. They are hot but they are small, so according to the laws of radiation they are not that luminous.

> **Q21. Stars with masses between 0.4 to 8 solar masses end their lives**
>
> a. with a violent explosion or supernova
>
> b. collapsing into a hot object known as a neutron star
>
> c. with a gentle but persistent stellar wind that destroys the entire star, leaving nothing behind
>
> d. with a gentle but persistent stellar wind that destroys the envelope of the stars leaving behind a white dwarf
>
> e. at the end of the hydrogen cycle

Q22. The refusal of the electrons to be packed closer together in a white dwarf produces the _____ that balances the gravitational force.

a. outward force of pressure
b. heat
c. force of attraction
d. spin

Q23. The minimum mass for an object to initiate thermonuclear fusion is about _____.

a. 2 solar masses
b. twice the mass of Jupiter
c. 0.8 solar masses
d. 0.08 solar masses

Q24. Most of the white dwarfs imaged with telescopes are close to the Sun because _____.

a. they form only close to the Sun
b. they emit only in the infrared
c. they are made of low density cold hydrogen
d. they are not that luminous and cannot be seen at great distances
e. a and b

In chapter 7, we saw that globular clusters are more than 10 billion years old and, white dwarfs are members of this clusters.

White dwarfs are not luminous, so they are difficult to imige and only the ones near the Sun can be observed.

White dwarfs, are members of a group of objects called **compact objects**.

White dwarfs are very surprising compact objects. For example, if you add mass to a white dwarf that has a smaller mass than the mass limit the dwarf shrinks, because more mass means more gravitational pull. Thus, a 0.9 solar mass white dwarf is bigger than a 1.4 solar mass white dwarfs. (Does this make sense to you?). We have more surprises coming related to white dwarfs.

Interacting Binary System and White Dwarfs

We know that the majority of the stars are "binary stars," so we expect many white dwarfs to have a companion. When the binaries are located at large distances from each other, they do not interact, but when the separation is small (no more than a few AUs), gas can flow from the red giant to a rapidly rotating whirl pool, or accretion disk, around the white dwarf from where it falls to the surface of the dwarf.

The gas in the accretion disk rotates fast and is very hot. The gas rotates fast because the gas particles in the accretion disk move in a smaller orbit than when they were in the atmosphere of the red giant (conservation of angular momentum law). The gas is hot because it has fallen a great distance, from the upper atmosphere of the giant to the accretion disk of the white dwarf (conversion of gravitational potential energy into heat). The gas is also hot because of the friction between the particles in the fast rotating disk. The inner layers of the accretion disk are hotter than the outer layers because the gas particles have travelled longer, and often they emit X-ray radiation. These systems are known as **X-ray binaries**.

There are two interesting cases that result from the flow of mass from a red giant to the surface of the white dwarf: Nova and supernova type Ia.

Nova

The gas that accumulates on the surface of the dwarf forms a small, thin, hot, and highly compressed hydrogen atmosphere that suddenly fuses into helium. This sudden and violent explosion blows away the entire atmosphere around the dwarf, leaving its surface intact. This event is known as **nova**. The explosion causes the luminosity of the dwarf to increase rapidly and then gradually fades away as shown in the light curve of Figure 8-10.

At their maximum, the novae are about 10,000 times more luminous than the Sun. Therefore, they can be seen from far away.

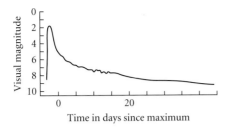

FIGURE 8-10. *Light curve Nova Cygni 1975.*

In fact, our modern telescopes routinely observe novae, and many have been seen with the naked eye when they were at their brightest. On average, two novae are observed every year, in our galaxy and other galaxies Galaxy.

These events are called **novae**, meaning new, because they suddenly appear in the sky and slowly fade away. In Figure 8-11, you can see the same region of the space before and after the appearance of a nova.

The explosion leaves the white dwarf intact and, as long as the companion dumps gas on the dwarf, the process repeats with intervals of a few decades, hundreds or thousands of years, or tens of thousands of years. This is known as **recurrent novae**.

The light curve of different novae events are similar.

Supernova Type Ia

If the gas that accumulates on the dwarf causes its mass to exceed the Chandrasekhar limit of 1.4 solar mass, electron degeneracy pressure is not able to balance the gravitational force. The dwarf then shrinks very fast and experiences rapid increase in its temperature. When the temperature nears 600 million K, the inert carbon that forms the bulk of the dwarf explosively begins the carbon cycle, which blasts away the entire dwarf in what is known as **supernova type Ia**. (Later on in this chapter, we will learn that a high mass star produces also a supernova that is known as **supernova type II**.)

The violent fusion of carbon gives rise to the formation of radioactive nickel that decays into radioactive cobalt and cobalt in turn decays into iron.

The explosion ejects most of the mass of the dwarf in the form of iron, along with other heavier elements, that form right after the explosion. These elements mingle with the interstellar medium, enriching it with heavier elements than helium. About one-third of the iron in the Milky Way is made in these type of events.

Immediately after the explosion, the luminosity of a supernova suddenly increases to more than one billion times the luminosity of the Sun. It can outshine the luminosity of the galaxy where it occurs.

Q25. Which of the following sequences correctly describes the stages of life for a medium mass star?

a. protostar, red giant, main sequence, white dwarf

b. protostar, main sequence, white dwarf, red giant

c. white dwarf, main sequence, red giant, protostar

d. red giant, protostar, main sequence, white dwarf

e. protostar, main sequence, red giant, white dwarf

Q26. What happens to the core of a Sun-like star after a planetary nebula occurs?

a. It breaks apart in a violent explosion

b. It becomes a neutron star

c. It contracts from a protostar to a main sequence star

d. It becomes a white dwarf

Q27. White dwarfs and the cores of some red giants support their weight by the pressure produced by the

a. refusal of the electrons to be packed closer together

b. fast expansion of the core

c. heat liberated in the fusion of carbon into silicon

d. heat liberated by the fusion of iron

The brightness of a typical Type Ia supernova reaches the maximum in about 20 days after the explosion and gradually fades away in about 40 days. See Figure 8-11.

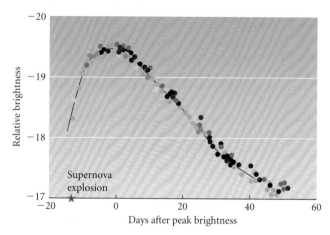

FIGURE 8-11. *Light curve of a supernova type Ia. Notice that at a maximum the absolute magnitude is about 19.5.*

Type Ia supernovae are always produced by white dwarfs nearing the 1.4 solar mass mark, the Chandrasekhar limit. Therefore, they always show the same characteristics, independent of the time and place where they are produced. For example, all of type Ia supernovae display the same maxima in their luminosity and the same absolute magnitude of about −19.4. See Figure 8-11. For this reason, astronomers use them to study the structure and evolution of the universe and as "yardsticks" for calculating the distance to the host galaxies where the explosion is observed.

The explosion destroys the white dwarf and produces a shell of expanding gases or supernova remnants that can be seen many years after the explosion.

The absorption spectra of type Ia supernova, is characterized by the presence of silicon II lines at 615 nm. In addition, the absorption line also has sulfur, calcium, iron, and nickel-56. Some of these elements were formed prior to the explosion, and others right after the explosion. The spectra lack hydrogen lines because the dwarf does not contain much hydrogen. See Figure 8-12.

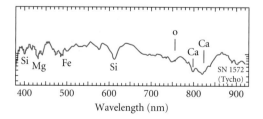

FIGURE 8-12. *The absorption spectrum of a supernova type Ia event lacks hydrogen lines.*

Postmain Sequence Evolution of a High Mass Star

High mass main sequence stars have more than 8 solar masses. These stars are located in the upper left-hand corner in the H–R diagram, and are hot, big, and luminous. See Figure 8-13. Because of main sequence their large mass, they remain as main sequence stars for a short time compared with lower mass stars.

Let's follow the evolution of a postmain sequence high mass star of 15 solar masses. This high mass star, like a medium mass star, also fuses hydrogen into helium in its core but only for about 11 million years. At the end of the hydrogen cycle, the core contracts and heats up, its envelope expands and cools down. The increase in temperature in the core causes hydrogen to fuse in the shell and the star enters in the super giant branch. See Figure 8-13. (Compare the evolution track with the one described in Figure 9-6 for a medium mass star.)

The surface temperature of this high mass main sequence star is about 30,000 K, given to the star its bluish color. Therefore, when it leaves the main sequence, its surface temperature drops, it changes color from blue to yellow, and the star becomes a yellow super giant. See in Figure 8-13. Even though the surface temperature of the giant has dropped, its luminosity remains almost constant because its volume increases.

At the end of the hydrogen cycle, the core contracts and heats up. The temperature easily reaches the 100 million K needed to initiate the helium cycle without the helium flash. When the helium in the core is exhausted, the core contracts and gets hotter and the star begins to fuse helium in the shell. The envelope expands and cools down becoming an even larger yellow supergiant. At this point in the evolution, the luminosity of these yellow super giant changes rhythmically, as was the case for the yellow giants arising from medium mass stars. But these giants are much hotter and more luminous. Therefore, the period of pulsation is longer. These yellow variables are known as **yellow Cepheids variables**.

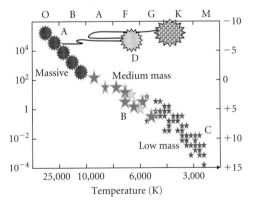

FIGURE 8-13. *The luminosity of super gaint remains fairly constant during their evolution from main sequence stars.*

Q28. The period of a yellow giant variable star is 40 days. This variable star probably is _____.

a. Planetary nebula
b. RR-Lyrae
c. T-Tauri
d. Cepheid

Q29. The absorption spectrum of a supernova type Ia is characterized by the absence of _____. Hint: See Figure 8-12.

a. calcium
b. silicon II
c. hydrogen
d. iron

Q30. The period of the Cepheid whose light curve is given in Figure 8-21 is _____ days.

a. 4
b. 6
c. 14
d. 20

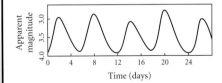

FIGURE 8-21. *Light curve of a yellow Cepheid variable star for problem 30.*

Chapter 8—Death of the Star

Figure 8-14 shows the light curve of a yellow Cepheid whose apparent brightness oscillates from a minimum of 25.6 to a maximum of 24.6. in about 50 days.

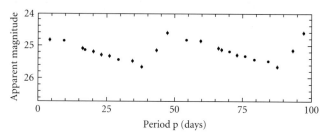

FIGURE 8-14. *Light curve for a Cepheid variable.*

Note that the light curve of the Cepheid is asymmetrical, showing that the time from dimmest to brightest is shorter than from brightest to dimmest.

In general, the period of oscillation of yellow Cepheids is larger than for RR-Lyrae. Some Cepheids have periods of less than 20 days and others have upto 80 days.

We said earlier that yellow Cepheid variables are going through the helium cycle. In other words, they are building a carbon-oxygen core. At the end of the helium cycle, the carbon-oxygen core contracts and grows hotter. When the core reaches a temperature of 600 million K, ignition of carbon in the core begins. When the carbon in the core is exhausted, the star will begin to fuse carbon in the shell.

The fusion of carbon produces neon, magnesium, silicon, and finally iron. After each element is depleted, the core contracts and heats up. Then the fusion of that element begins in the shell. This pattern of **core fusion** and **shell fusion** continues for each element, and the star develops a layer structure as shown in Figure 8-15. The last element to fuse is silicon into iron.

The fusion of each element gets faster and faster at a consecutively higher temperature, as shown in Table 8-1, for the case of a 25 solar mass star. The table shows that hydrogen fuses 7 million years while Silicon fuses in only 1 day.

> **Q31.** The apparent magnitude when the Cepheid variable of Figure 8-22 looks the brightest is
>
> a. 16.8
> b. 16.6
> c. 17.0
>
> **Q32.** The apparent magnitude when the Cepheid variable of Figure 8-22 looks the dimmest is
>
> a. 16.6
> b. 16.8
> c. 17
>
> **Q33.** The average apparent magnitude of the Cepheid variable of Figure 8-22 is
>
> a. 16.6
> b. 16.8
> c. 17
>
>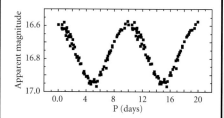
>
> **FIGURE 8-22.** *Light curve of yellow Cepheid variable star for problem 31–33.*

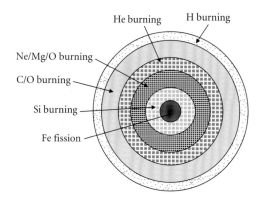

FIGURE 8-15. *Shells burning inside massive stars as they evolve from main sequence star. The last element to fuse is silicon (Si) into iron (Fe).*

Chapter 8—*Death of the Star*

TABLE 8-1. *Heavy element fusion in a star of 25 solar mass.*

FUEL	TEMPERATURE (K)	TIME TO NEXT CYCLE
H	29 million	7 million years
He	100 million	0.5 million years
C	600 million	600 years
Ne	1 billion	1 year
Si	2 billion	1 day
Fe core collapse	3 billion	0.1 s

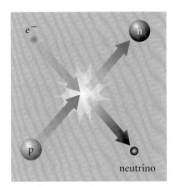

FIGURE 8-16. *When the core of a massive star collapses, protons recombine with electrons to form neutrons liberating neutrinos.*

The core of high mass stars is a factory which produces elements from helium to iron. This process is known as **nucleosynthesis**.

By the time the star fuses silicon into iron, the core has shrunk to a diameter the size of the Earth and has the incredible temperature of approximately 2 billion K.

The formation of iron in the core changes everything in the dying star. Because iron is the most stable element in the universe, it does not fuse into heavier elements. Instead, iron undergoes fission, i.e., it breaks into lighter elements and ultimately breaks into **protons, neutrons**, and **neutrinos**. This process is known as the **photodisintegration** of iron. See Figure 8-16. Fission needs energy. So the fission of iron—instead of making energy—removes energy from the core, lowering its temperature and pressure and ultimately causes it to collapse.

In less than a second, an iron core as big as the Earth collapses into a small object of about 10 to 20 miles in diameter. As the core collapses, the photodisintegrated iron protons capture free electrons to form neutrons. These will pile up in a ball of neutrons. The final compact object formed is called a **neutrons**. The electrons captured by the protons also liberates an avalanche of neutrinos that flow outwardly along with the huge shock wave that arises after the collapse of the core.

> **Q34. The last element the stars form by nucleosynthesis probably is**
>
> a. iron
> b. silicon
> c. carbon
> d. calcium
>
> **Q35. Most of the energy produced during a supernova explosion is carried away in the form of**
>
> a. visible photons
> b. positrons
> c. gamma ray
> d. X-ray
> e. neutrinos

As the core collapses, the star's atmosphere loses its support and rushes down toward the center of the star. The outward shock wave, along with the flow of neutrinos, meet the falling atmosphere, producing a great explosion in which the atmosphere of the super giant is blown away. This is known as a **supernova type II**.

During the explosion, fast nuclear reactions are set on and elements heavier than iron are produced, such as calcium, iodine, gold, and platinum.

The explosion spreads the newly formed nuclei out into the interstellar space, along with the nuclei created by normal nucleosynthes in the core of the star. Supernovae enriches the interstellar matter with elements heavier than helium. These elements are incorporated into a new generation of stars and new planets. Our body carry elements that were formed in dying stars.

The neutrinos produced in the photodisintegration of the iron nuclei carry about 99% of the energy produced in the explosion.

The death of massive stars is known as **core collapse supernova**.

The most massive stars, with the shortest lifetimes and the most energetic supernovae explosions, collapse into an even smaller, denser core from which not even light can escape. This is a black hole.

Supernovae are cataclysmic stellar explosions that are accompanied by bright visual displays. They briefly produce as much energy as the entire galaxy where they are located.

Figure 8-17 shows the light curve for supernova type Ia and type II. The vertical axis gives the absolute magnitude instead of the luminosity.

Recalling that there is a correspondence between absolute magnitude and luminosity, the graph of Figure 8-17 indicates that the corresponding luminosity values are about a billion

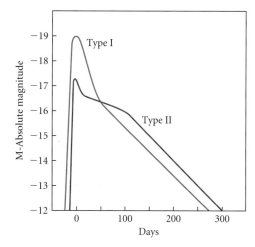

FIGURE 8-17. *Comparison of the light curve of supernovae type Ia and type II. Note that type Ia is brighter.*

170   Chapter 8—*Death of the Star*

solar masses at their maxima. Notice that type Ia is brighter than type II.

The material ejected by the explosion of the supernova moves outwardly at a speed of approximately 20,000 km/s, carrying most of the mass of the exploding stars and all the elements made by nucleosynthesis and those made during the supernova explosion. This mass of expanding nebula is known as **supernova remnant**.

As the years pass, the supernova remnants become thinner and fade, mingling with and enriching the interstellar matter.

Both type of supernovae produce a supernova remnant. This nebula is a powerful source of radio emission called **synchrotron radiation** that will be explained in the next chapter.

The absorption spectrum of a supernova type II has hydrogen lines because the explosion involves the atmosphere of the dying stars that are rich in hydrogen. See Figure 8-18.

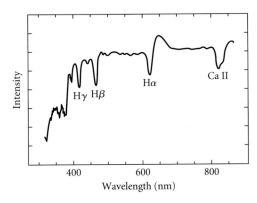

FIGURE 8-18. *Absorption spectrum of the 1987 supernova. Note that the spectrum contains hydrogen lines.*

Supernovae type I are observed in binary systems, usually in globular clusters. Supernovae type II are observed in open clusters because they are produced by high mass stars, and high mass stars are always members of young open clusters.

Supernovae remnants and planetary nebulae are good evidence that stars dye.

Supernovae are rare—they happen about every 1,000 years in our galaxy. Only a few have been seen with the naked eye. Arabic astronomers saw one in 1006, the Chinese astronomers saw another in 1054, Tycho Brahe saw one in 1572, and Kepler saw one in 1604.

On February 23, 1987, 19 neutrinos coming from the Large Magellanic Cloud, a small satellite galaxy to the Milky Way, were detected. When the telescopes were directed to the region where the neutrinos were formed, a supernova explosion was discovered. For some time, the supernova was visible to the unaided eye. Astronomers have been following the evolution of the supernova and the remnants of the supernova since the explosion was discovered.

Q36. How does the life of a high mass star differ from the Sun's life?

a. It forms much faster

b. It lives a shorter time on the main sequence

c. As a super giant it makes elements heavier than carbon

d. When it dies, it explodes in to a tremendous supernova explosion

e. All of the above

Q37. Figure 8-23a shows the light curve of _____ and Figure 8-23b the light curve of _____.

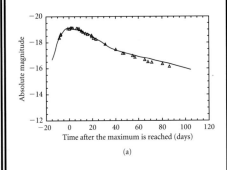

(a)

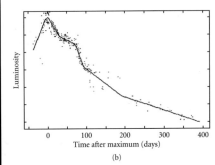

(b)

FIGURE 8-23. *For problem 37.* http://adsabs.harvard.edu/abs/1985AJ.....90.2303D.

Chapter 8—*Death of the Star*

The supernova that was detected in 1987 has provided valuable information about the chemical composition and evolution of supernovae type II events.

CEPHEID VARIABLES STARS AS DISTANCE INDICATORS

The Cepheids variables are important because their periods of variability are related to their absolute magnitude or luminosity, as shown in Figures 8-19 and 8-20.

Therefore, by measuring the period of the Cepheid variables, the luminosity and absolute magnitude can be calculated. The apparent magnitude of the Cepheid is measured easily with a telescope. Therefore, if we know the apparent and absolute magnitudes, the distance in parsecs (pc) to the Cepheid can be obtained using the well-known formula as follows:

$$d = 10^{\frac{m - M + 5}{5}}$$

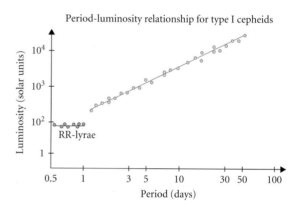

FIGURE 8-19. *Luminoity period relation for RR-Lyrae and yellow Cepheids variable stars.*

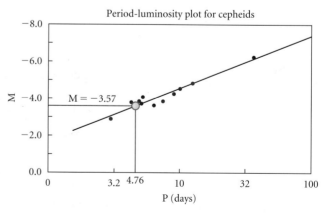

FIGURE 8-20. *Period absolute magnitude curve for Cepheid variables.*

> **Q38. Use Figure 8-24 to answer the following questions.**
>
> a. Name four red giants
>
> b. Name two blue supergiants
>
> c. Name four main sequence stars
>
> d. Name two white dwarfs
>
>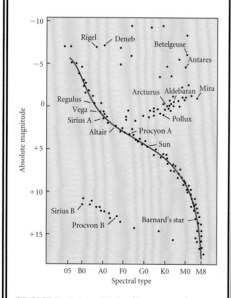
>
> **FIGURE 8-24.** *H–R diagram of some well-known stars for problem 38.*

Chapter 8—Death of the Star

Example

The apparent magnitude of the Cepheid marked with a circle, in Figure 8-19 is 11.43. How far is the Cepheid?

Solution Using the previous equation, we have

$$d = 10^{\frac{11.43-(-3.57)+5}{5}} = 10^4 \text{ pc}$$

ANSWERS

1. d
2. c
3. a
4. b
5. d
6. c
7. b
8. a
9. d
10. c
11. d
12. d
13. b
14. c
15. b, Measure the distance (time difference) between two consecutive minima or maxima. A more accurate value is obtained if you use time difference between the last and first minima and divide the result by four: (2.7 − 0.4 = 2.3 days/4 = 0.56 days) or about 0.6 days.
16. c
17. All RR-Lyrae have the same absolute magnitude so all we need to measure is the apparent magnitude and use equation 6-1
18. b
19. d
20. b
21. d
22. a
23. d
24. d
25. e
26. d
27. a
28. d
29. c
30. b
31. b
32. c
33. b
34. a
35. e
36. e
37. a, Supernova type Ia. b, supernova type II. Notice that the light curve of Type II has a plateau.
38. a. Mira, Aldebaran, Arcturus, Pollux
    b. Rigel, Deneb
    c. Barnard's star, Altair, Vega, Regulus
    d. Sirius B, Procyon B

CHAPTER 9

Neutron Stars and Black Holes

In chapters six and seven, we saw how the stars are born, evolve, and finally pass away. The manner of their passing away and what they leave behind depends on the star's mass. Medium mass stars end their lives with a gentle wind called a planetary nebula, and they leave behind a white dwarf. High mass stars die with a loud explosion, or supernova type II, leaving behind a neutron star or a black hole.

We have already studied white dwarfs. In this chapter, we will concentrate on the objects produced when high mass stars die.

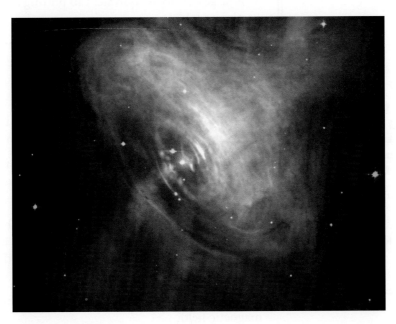

FIGURE 9-1. *The heart of the Crab Nebula.*

Credit: NASA Hubble Space Telescope Collection

Neutron Stars

In 1934, Walter Baade and Fritz Zwicky suggested that neutron stars (NS) were formed during a supernova core collapse. Most astronomers ignored this theory about NS because these objects were very unlikely to be observed. Today, we have abundant experimental observations that support Bade and Zwicky's theory. The first experimental observations for their existence were provided in 1967.

In the previous chapter, we saw that when a high mass star builds an iron core, the core collapses in less than a second into a compact object. If the collapsing core of the dying high mass star has a mass between 1.4 and 3 solar masses, a NS is produced, and if the mass is larger than three solar masses, the core collapses into a black hole.

First, we will study NS.

NS are formed primarily by neutrons with a few protons and electrons on the surface.

In chapter six, we saw that white dwarfs are supported against gravitational pull by electron degeneracy pressure. In the case of NS, they are supported by neutron degeneracy pressure. So NS are held by neutron degeneracy. Loosely speaking, the refusal of the neutrons to get closer together provides the force of pressure that balances the gravitational force in these compact objects. Neutron degeneracy established the maximum limit that NS can have. Theory predicts that NS have a mass between 1.4 and 3 solar masses, concentrated in a small volume of a 10- to 20-km diameter, which gives a density of about 10^{14} g/cm^3.

This large density is equivalent to putting the six billion people of Earth in a tablespoon.

This large concentration of mass in such a small volume produces a strong gravitational field, about 10^{11} times the Earth's gravity. If you were on the surface of one of these objects, you would be squeezed paper thin, and you would weigh more than a million tons!

One consequence of the high mass and the large density of NS is their huge escape speed, close to 150,000 km per second.

Astronomers believe that main sequence stars with mass between 8 and 25 solar masses will eventually produce a NS. If the initial mass is larger, they will produce a black hole.

NS rotate extremely fast, some as fast as 330 times per second, or even faster. Many NS have strong, rotating magnetic fields. The intensity of these fields is about one billion times the value of the magnetic field on the Sun.

Why do NS rotate that fast, and why do they have such a large magnetic field?

Q1. NS are produced when medium mass stars like our Sun die in the form of a planetary nebula.

a. True
b. False

Q2. A new NS was detected inside a supernova remnant with a mass of about 4.5 solar masses.

a. True
b. False

Q3. NS have more mass than white dwarfs and, therefore, they are bigger.

a. True
b. False

Q4. NS are supported against gravitational collapse by pressure degeneracy produced by degenerate _____.

a. protons
b. electrons
c. neutrons
d. photons
e. neutrinos

Q5. Which of the following objects has the largest gravitational force on its surface?

a. The Earth
b. A brown dwarf
c. A white dwarf
d. A NS

As the core of the star collapses, any angular rotation of the core will be amplified, as required by the conservation of angular momentum law, causing the NS to rotate fast. Similarly, the original magnetic field of the parent star would be squeezed into the small volume of the NS, thereby creating a very intense gravitational field on the collapsed core.

Other theories propose that the magnetic field that the NS display might have formed during the collapse of the core of a dying star that gave rise to the NS.

Because NS result from the core collapse of a high mass star, their initial temperatures are close to 100 billion K, and with time, they cool down to about one million K.

Some NS are inside supernovae remnants, but others move through space at great speeds. These moving NS probably received a kicked during the explosion of the massive star that gave origin to them.

The Crab Nebula is the remnant of a supernova that exploded in 1054 AD, and that was observed by Chinese astronomers. See Figure 9-2. It contains a rapidly rotating NS in its interior.

NS are very small, and although they are hot, they are not luminous, so it is very difficult to detect them directly with an optical telescope. So, how do we know that NS exist?

The answer is pulsars.

> **Q6. Which of the flowing is not a characteristic of a NS?**
> a. Large density
> b. Fast rotation
> c. A strong magnetic field
> d. As big as the Earth
>
> **Q7. Newly formed NS have powerful magnetic fields.**
> a. True
> b. False

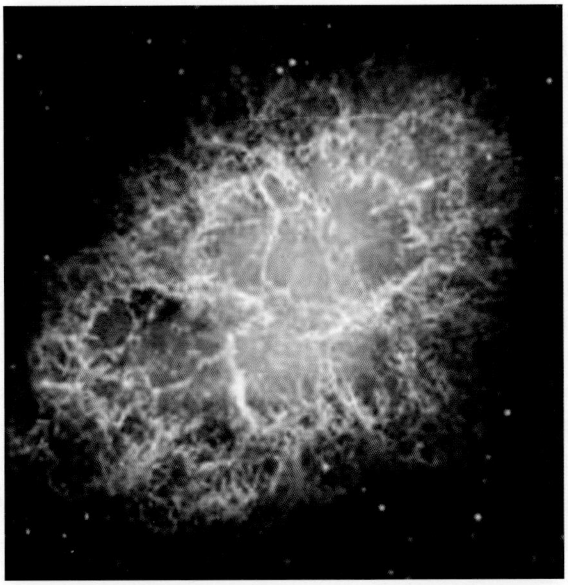

FIGURE 9-2. *It is believed that at the center of the Crab Nebula there is a fast-rotating neutron star.*

Credit: NASA Chandra Space Telescope Collection

Chapter 9—Neutron Stars and Black Holes

Q8. What is the period of a NS that rotates 25 times per second?

a. 2.5 s
b. 0.05 s
c. 0.025 s
d. 0.004 s
e. 0.04 s

Q9. The frequency of the pulsar of Figure 9-3 is close to _____ revolutions per second.

a. 8
b. 2.04
c. 0.125
d. 2.5
e. 1.25

Q10. Pulsation of normal stars cannot be the cause of pulsars because the luminosity of the stars can not change hundreds of time per second.

a. True
b. False

PULSARS

Pulsars are fast-rotating NS that emit a bright pulse of radiation toward us each time they rotate once. These pulses are most easily detected with radio telescopes.

Pulsars were discovered in 1967 by Jocelyn Bell Burnell, a 24-year-old graduate student who was doing research with a radio telescope. She detected a regular radio pulse with a period of 1 1/3 s. Initially, there was no valid physical explanation for the source of this intriguing periodic radio signal, and it was attributed to an extraterrestrial civilization that was called **LGM**, for little green men. Soon after Jocelyn's discovery, more radio pulses were found with periods of 0.4–4 s. This forced astronomers to look for a more convincing explanation.

They called the objects that generated these pulses as **pulsars**.

What is the source of this radio frequency radiation?

Because the signal received is discrete and periodic, the period (P) of the source of the radiation must be equal to the period of the pulses detected. For example, the period of the pulse shown in Figure 9-3 is 0.8 s, and therefore that is the period of change of the radiation that produces the pulse.

Several mechanisms were considered as the source of radiation that produces the pulse. The mechanisms include pulsation, orbital motion, and rotation associated with normal stars, white dwarfs or NS.

Pulsation of normal stars was quickly ruled out because the luminosity of the stars cannot change in such a short period (a few seconds). We have already seen that variable stars have periods of days or hours. White dwarfs are solid objects and cannot contract and expand that fast. Perhaps the same is true for NS.

The orbital motion of stars, white dwarfs, and NS around a common center of mass was also analyzed, but none of these provided a satisfactory explanation for the cause of the pulsated radiation.

Finally the third mechanism, or rotation, was considered. The object that gives rise to radio signals with such a short period had to be spinning very fast, more than 100 times per second. This rapid rotation excluded normal stars and white dwarfs as the source of the signal because they cannot rotate with such high frequency. Can you image a white dwarf or a star rotating more than 100 times per second? They would be ripped apart. But NS, being small, can rotate very fast.

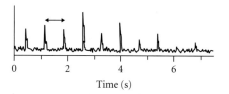

FIGURE 9-3. *The period of this pulse is about 0.8 s.*

The mechanism that gives rise to the pulses has been linked to the magnetic field associated with the NS and can be explained as follows:

Close to the magnetic poles, a NS has two "hot spots" of charged particles, mainly protons and electrons. The high speed of rotation of the NS accelerates these particles to extremely high speeds along the magnetic field axis. As the charged particles spiral along the magnetic field axis, they emit a beam of radiation. The radiation produced in this manner is called **synchrotron radiation**, and it is most easily detected with radio telescopes.

It happens that the magnetic field axis is inclined with respect to the axis of rotation of the NS, as shown in Figure 9-4. So as the NS rotates, the magnetic field axis and the radiated beam also rotate. Because of the tilt of the magnetic axis, the radiation emitted along this axis sweeps a cone around the axis of rotation. See Figure 9-5. If the Earth is in the path of the beam, every time the beam sweeps in front of the Earth a pulse of energy is detected.

The NS is not pulsing on and off; we see a pulse as a beam sweeps in front of our telescopes. See Figure 9-5a. A pulsar is like a lighthouse with a rotating light beam (**the lighthouse model**).

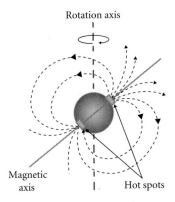

FIGURE 9-4. *The magnetic and rotation axis are not aligned in neutron stars. The broken lines represent the direction of the magnetic field lines.*

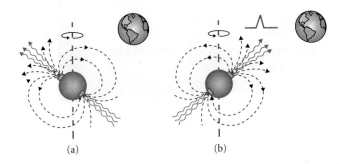

FIGURE 9-5. *(a) When the cone of radiation points away from Earth, no signal is detected. (b) When the cone flashes in front of the Earth, a signal is picked up.*

Summarizing, the lighthouse model requires the following:

1. A spinning NS that emits a continuous radiation along its magnetic axis.
2. A small tilt between the magnetic axis and the rotation axis.
3. A beam sweeping past the Earth.

As a result, we can see a pulse.

Most pulsars have periods between 0.03 and 0.3 s (or 33 to 3.3 revolutions per second) and are sources of intense radio radiation. A few also emit X-rays, gamma rays, and visible radiation.

Recall that the rotation period (P) of an NS equals the period of the pulses of the emitted radiation.

Many NS and pulsars are inside supernovae remnants.

About 1,500 pulsars have been found so far.

At the heart of the Crab Nebula lies a young and powerful pulsar that emits X-rays, visible rays, and infrared and radio radiations. The NS that gives rise to the Crab Pulsar rotates about 30 times per second, corresponding to a period of 0.033 s.

The energy emitted by the pulsar ionizes the supernova remnants, making them glow as shown in Figure 9-6a. The Crab Pulsar is only 900 years old. It is the youngest pulsar ever detected.

Not all supernova remnants have pulsars inside because pulsars remain detectable for about 10 million years, and supernova remnants cannot survive for more than 50,000 years. Also there are NS and pulsars moving through space at high velocities because they received a kick when the NS was formed.

Observations have shown that the duration of the pulse of pulsars slowly increases with time, implying that the rotation of the NS that produce the pulsar is slowing down. This should not surprise us because NS are converting rotation energy into radiation energy. Consequently, young pulsars have short periods because they spin fast, and old pulsars have longer periods because they rotate slowly.

Q11. NS have two magnetic poles (N and S) and they therefore behave as large magnets. The magnetic axis joining the two poles _____.

a. lies along the axis of rotation
b. lies perpendicular to the axis of rotation
c. makes a small angle with the axis of rotation
d. constantly changes its direction with respect to the axis of rotation

Q12. All NS are detected as pulsars.

a. True
b. False

Q13. The NS that produces a pulsar rotates 100 times per second. The period of the pulse in seconds is _____.

a. 1,000
b. 10
c. 0.1
d. 0.01
e. 0.001

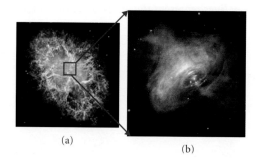

(a) (b)

FIGURE 9-6. (a) *The Crab Nebula. In the center of the square, a neutron star is visible.* (b) *A composite image of the Crab Nebula showing the X-ray (blue) and optical (red) images superimposed.*

Credit: (a) NASA Hubble Space Telescope Collection; (b) NASA Chandra Space Telescope Collection

Neutron Stars in a Binary System

Besides radio pulsars, astronomers have also detected X-ray pulsars associated with a binary system formed by a NS and a companion star.

Often the gas from the companion star forms a hot, fast-rotating accretion disk around the NS. The inner layer of the accretion disk is hotter than the outer layers because the gas particles have fallen through a larger distance from the atmosphere of the companion, and thus they have gained more kinetic energy. More kinetic energy means a higher temperature.

The gas in these inner regions is so hot that it emits continuous X-ray radiation.

The hydrogen that falls onto the surface of the NS, besides being hot, is highly compressed and fuses into helium. When the helium reaches a temperature above 100 million K, it suddenly ignites over the entire body of the NS, producing a large burst of X-ray radiation energy, lasting only few seconds. See Figure 9-7.

These X-ray emissions are called **X-ray bursters**. As long as gas keeps flowing from the companion star, the bursts reappear every few hours or so.

The bursts of X-ray energy are superimposed on the continuous X-ray radiation described earlier. Notice the similarity between the X-ray bursts and the nova explosions described in chapter eight. The surface of the NS is not affected by the bursts.

In other cases, as the gas from the companion falls toward the NS, its powerful magnetic field directs the incoming matter to the neighborhood of the magnetic poles, forming two hot spots near the magnetic poles.

These hot spots have temperatures of about 100 million K, and they emit two beams of X-ray radiation. As the NS rotates on its axis, the beams sweep the sky, and if the Earth is in the path of

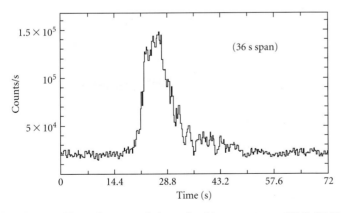

FIGURE 9-7. *X-ray burst emission of a binary system GRO 1744-28, involving a neutron star.*

one of them, a pulsating X-ray signal is detected (lighthouse model). As before, the frequency of the X-ray pulse tells us the frequency of rotation of the NS.

X-ray eclipsing binaries are also produced by these systems. The Hercules X-1, located about 15,000 ly away in the constellation Hercules, is a well-known eclipsing binary. This system emits an X-ray pulse every 1.2378 s, but every 1.7 days, it disappears for 0.24 days (5.76 h), producing an eclipse, as the NS goes behind the companion star.

Hundreds of X-ray binaries have been detected in our galaxy, including X-ray eclipsing binaries.

The pulsar PSR 1931+16 is an intriguing radio pulsar. Every 7 h and 45 min, the length of its pulse, grows alternatively longer and then shorter. To explain this anomalous behavior, it has been assumed that the NS that generates the radio pulse orbits another NS in a small orbit. When the NS moves in its orbit away from Earth, the radiation is redshifted, and the period gets longer, and when it moves toward the Earth, the radiation is blueshifted, and the period of radiation grows shorter.

Another type of pulsar associated with binary systems is the millisecond pulsar.

Millisecond Pulsar

Millisecond pulsars rotate very fast and have pulses of millisecond duration. For example, if the NS of a pulsar rotates 625 times per second, it will produce a pulse with a period of P (P = 1/f = 1/625 = 0.0016 s) of 1.6 milliseconds.

Millisecond pulsars are the fastest known pulsars, and they are always members of a binary system. What is strange about millisecond pulsars is that they involve old NS.

(Remember that only young pulsars pulsate fast.)

This is possible only if the NS is a member of a binary system in which the companion is a normal star or a giant star. As mass from the atmosphere of the companion star falls to the surface of the NS via the accretion disk, angular momentum is transferred to the NS, causing it to spin faster. The NS has been rejuvenated by recycling matter from the companion.

About 150 millisecond pulsars have been discovered, and many of them are in **globular clusters**.

Why are most of the millisecond pulsars discovered in globular clusters?

Escape Speed in Compact Objects and Black Holes

The value of the escape speed from the surface of an object depends on the mass, **M**, and the radius, **R**, of the object, we are trying to escape from. Using elementary Physics it can be shown that the value of the escape speed is obtained using the following equation:

$$V_{escape} = \sqrt{\frac{2GM}{R}}, \qquad (9\text{-}1)$$

where **G** is the constant of universal gravitation.

Using equation 9-1, we find that the escape speed from the surface of the Earth is only 11 km/s, from the Sun 600 km/s, from the surface of a white dwarf about 6,000 km/s, and from the surface of a NS close to 150,000 km/s, or 50% of the speed of light.

The trend is clear. The smaller the volume in which a mass is concentrated, the larger its escape speed, and the harder it is to escape from its surface. The following question comes to mind. Is there a place in the universe where the escape speed is equal to the speed of light?

The answer is yes. And this creates a big problem, because matter cannot travel at the speed of light. Therefore, nothing will escape from that region. To an outside observer this region would look black, and we call it a **black hole**. A black hole is an object whose escape speed near its surface exceeds the speed of light. Traditionally, this surface is called the **event horizon** because we are unable to observe anything inside it.

Black holes are completely inaccessible; they are completely isolated from the rest of the universe because nothing comes out of their event horizon. We have no way of knowing what is inside the event horizon of a black hole. The radius of the event horizon of a black hole is called the **Schwarzchild radius (Rs)**. See Figure 9-8.

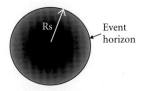

FIGURE 9-8. *No information goes out through the event horizon of a black hole.*

Q14. From which of the following objects is the escape speed the largest?

  a. The Earth
  b. A brown dwarf
  c. A white dwarf
  d. A NS
  e. A black hole

Q15. NS are smaller than white dwarfs, therefore the escape speed from NS is lower than the escape speed from white dwarfs.

  a. True
  b. False

Q16. The "surface" through which black holes interact with the rest of the universe is called the _____.

  a. Schwarzschild Radius, Rs
  b. Newtonian surface
  c. event horizon
  d. black surface

Q17. A black hole has a mass of 10 solar masses. What is the radius of the event horizon or Schwarzschild radius (Rs)?

  a. 10 km
  b. 10 m
  c. 30 km
  d. 20 km
  e. 3 km

To find the radius of the event horizon, we ask the following question. At what radius (**R**) does the escape speed from the event horizon of a black hole equal the speed of light (**c**)?

To find the answer, we replace the escape speed (**V**) in equation 9-1 with the speed of light (**c**) and solve for the radius (**R**), obtaining the following:

$$Rs = \frac{2GM}{c^2} \qquad (9\text{-}2)$$

The surface defined by this equation gives the radius of the event horizon, or **Rs**. If in this equation we replace the value of the speed of light and of the constant **G**, the radius of the event horizon takes a simple form:

$$Rs = 3M, \qquad (9\text{-}3)$$

where the radius is expressed in kilometer (km) and the mass in solar mass.

Question: What is the radius of the surface of the event horizon of a black hole of one solar mass?

Answer: Rs = 3M = 3 × 1 (km) = 3 km.

The Rs is the radius within which an object of mass (m) must shrink to become a black hole. For example, if the mass of the Sun is packed within a volume of 3 km of radius, it will form a black hole with an event horizon or Rs of 3 km.

Every object with mass has a Rs but not every object is a black hole.

If the Sun turns into a black hole, the gravitational force that it makes on the planets will not change, because the distance to the center of the black hole is exactly the same as the distance from the planets to the center of the Sun.

If we squeeze the Earth inside a volume of about 1 cm of radius, it would become a black hole with an event horizon, or Rs of 1 cm.

Gravity is able to prevent information or matter from coming out from inside black holes. Nothing, not even light or any type of electromagnetic radiation, escapes from a black hole. The gravitational force produced by a black hole is the only thing that is felt beyond the event horizon.

Black holes are better understood using Einstein's theory of general relativity. This theory predicts that black holes and, in general, any concentration of mass, distorts or curves the fabric of space-time. The more mass, the greater the distortion or curvature. In the jargon of relativistic physics, we say that the gravitational field produced by a mass warps or curves space-time. This curvature produces the gravitational attraction between the masses in the universe and changes the path and wavelengths of light as the light passes near an object with mass.

The warping of space is small if the masses are small, but large masses produce a large warp. Also at large distances from a mass, the warping of space (the gravitational force) is negligible.

All of us know that light travels in a straight light. This is true only in normal space where the gravitational field is weak or nonexistent. Where the space is curved, the light follows the space and no longer travels in a straight line. The effect is best observed when the light goes near a large mass, as is the case for a black hole, or for regions of space crowded with galaxies, as seen in Figure 9-9. In this figure, the light coming from star S seems to originate at S' (image) because the space-time around the mass is curved. This effect is called **gravitational lensing**, which we will discuss in our next chapter.

> **Q18.** A beam of light moving away from the event horizon of a black hole would _____.
>
> a. increase its speed
> b. be redshifted
> c. be blueshifted
> d. decrease its speed
> e. All of the above are true
>
> **Q19.** The wavelength of a beam of light moving away from the event horizon of a black hole _____.
>
> a. gets longer
> b. gets shorter
> c. remains unchanged

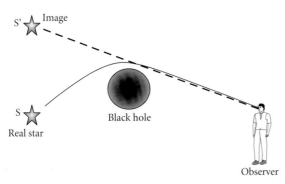

FIGURE 9-9. *The strong gravitational field of a black hole bends and changes the direction of light propagation.*

The curvature of space also changes the wavelength of light. The photons of light traveling in the direction of a gravitational field are blueshifted, and against it they are redshifted. Gravity attracts light. Therefore, a beam of photons of light traveling against a gravitational field must expend energy to escape gravity. But the photons of light cannot slow down because they are light, and light must travel at the speed of light, so the only alternative that the photons have to lose energy is to increase their wavelength or decrease their frequency. This effect is called **gravitational redshift**.

Similarly, a beam of photons of light traveling in the direction of a gravitational field would gain energy, causing its wavelength to get shorter, and therefore those photons would be **blueshifted**.

A beam of light pointed into space from inside a **black hole** would bend back toward its interior.

Figure 9-10 illustrates the effect on the light emitted by a star as it approaches a black hole. In (a), where the gravitational force from the black hole is weak, only a few rays are captured. In (b), the gravitational field is stronger than in (a), and the amount of light captured increases. The gravitational field is strong enough to

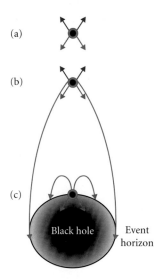

FIGURE 9-10. *The curved lines represent the light captured by the gravitational field of the black hole.*

bend the paths of the photons. When the star is near the event horizon (c), most of the radiation from the star is captured. And finally, when the star passes the event horizon, no radiation from the star escapes.

Warning: If you find a black hole, never go beyond the event horizon because you will never come back.

APPROACHING A BLACK HOLE

Imagine one of your friends leaving Earth toward a black hole in a spaceship and carrying an old-fashioned ticking clock and an ultraviolet distress signal light.

When the ship is far away from the black's hole event horizon, everything is normal in the ship because the force of gravity due to the black hole is practically zero.

But when the spaceship is close to the event horizon, the ship and everything in it is under a tremendous gravitational force that would destroy your friend and the ship, because the force that your friend's ship is experiencing is larger on the side facing the black hole than it is on the other side. The dust particles that remain from your friend and his ship would rain down to form a hot, swirling accretion disk around the black hole. Before the particles of the accretion disk crossed the event horizon, to their final destination in the black hole, they would emit an intense X-ray radiation into space.

But before your friend was destroyed, as he approached the black hole, the only thing that he would experience would be the increase of the gravitational force. He would see the clock ticking

> **Q20.** **When a beam of light passes close to the event horizon of a black hole, the beam _____.**
>
> a. keeps traveling in a straight line
>
> b. is deflected away from the black hole
>
> c. is bent towards the black hole

normally and the UV light emitting the right wavelength. But your friend's family on Earth, who were observing his journey with a telescope, would see a different picture. They would see the clock intervals getting longer, and the ultraviolet light from the ship being emitted at a longer wavelength the closer the ship was to the event horizon. First, they would observe violet wavelength radiation, then green, red, infrared, shortwave radio frequency, long wave radio frequency, and finally the wavelength gets so long that the ship simply vanishes, the clock stops, and they never see the ship crossing the event horizon. This change in the wavelength is due to the gravitational redshift because the light is traveling against the gravitational field of the black hole (See gravitational redshift on page 185).

The slowing down of the spaceship's clock is called **time dilation**, and it is closely related to the gravitational redshift. It is explained by the theory of relativity.

How do blacks holes form?

At the beginning of this chapter, we stated that if the mass of the core of a collapsing star is larger than three solar masses, the core would collapse into a black hole.

Theoretical calculations indicate that the cores of high mass stars with more than 25 solar mass would, at the end of their lives, collapse directly into a black hole. This event is called a **hypernova** or **collapsar**. These black holes are termed **stellar black holes**. Stellar black holes have only a few solar masses.

Some nonstellar black holes are produced by the collision of NS, and others seem to form at the center of galaxies.

The black holes at the center of the galaxies might have formed from matter that accumulated at the galactic center when the galaxies were forming and from collision between the stars that moved at great speeds near the center.

By measuring the speed of the dust and the speed of the stars near the galactic center, the mass of the black holes at the center of the galaxies can be calculated. These calculations show that the galactic central black holes have masses of several million solar masses. These black holes are called **supermassive black holes**.

Our own Milky Way seems to have a central supermassive black hole of about four million solar masses. (What is the radius of its event horizon?).

An observation that supports the idea of supermassive black holes at the center of the galaxies is the large amount of energy emitted, in different wavelengths, by tiny central galactic centers.

The only satisfactory explanation for the origin of this energy is to assume that the energy is produced and emitted by the accretion disks that form around the central galactic black holes. This is possible because the temperature of the gas in the accretion disk

Q21. A spaceship nearing the event horizon of a black hole is emitting an X-ray distress signal. You are in a mother ship at a safe distance from the event horizon. You would see the distress signal _____.

 a. in a shorter wavelength, as a gamma ray signal
 b. as an X-ray signal of the same wavelength
 c. in a longer wavelength, perhaps as infrared
 d. as an X-ray signal of shorter wavelength

Q22. If you were watching the collapse of the core of a high mass star directly into a black hole (a hypernova event), you would not be able to observe the formation of the black hole. The light disappears because _____.

 a. the color becomes black due to the fast contraction
 b. it is strongly blueshifted
 c. it is strongly redshifted
 d. you closed your eyes in the moment in which it collapsed

Chapter 9—Neutron Stars and Black Holes

reaches million of degrees and radiates energy in all the wavelengths of the electromagnetic spectrum. In fact, the so called active galaxies emit two powerful central beams of radiation opposite to each other, perpendicular to the accretion disk.

Gamma Ray Bursts

Hipernovae events are very violent and release tremendous amounts energy in the form of gamma ray bursts (GRB). A GRB is a sudden flare of gamma radiation, lasting only a few seconds as shown in Figure 9-11. GRB release more energy than X-ray bursts and last for less time.

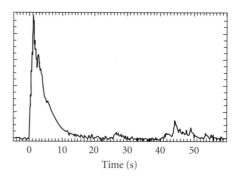

FIGURE 9-11. *Time profile of Trigger 3067 gamma ray bursts (GRB).*

GRB are also produced when two NS or two black holes merge.

The instruments on board of the Compton Gamma-Ray Observatory (CGRO), launched by NASA in the early 90s, detected about 2,500 GRB. See Figure 9-12. The gamma ray sources are uniformly distributed in the sky, and therefore they are produced far away in distant galaxies.

Currently, the NASA "Swift" observatory is dedicated to observing the events that involve the production of gamma energy in space.

The farthest gamma burst detected by the Swift project is the gamma burst GRB 080913, produced by a hipernova located 12.8 billion light years away. The information comes from near the edge of the visible universe.

Events that happen in the past have what astronomer call a "look back time." The look back time of GRB 080913 is 12.8 billion light years! At that time the universe was less than 825 billion years old. Our Sun has only 8.3 min of look back time.

Black Holes in Binary System

As in the cases of white dwarfs and NS, black holes can also be members of binary systems. When the companion of a black hole is a star, the strong gravitational field created by the black hole draws matter from the companion into an accretion disk outside and around the event horizon. In the accretion disk, the gas swirls,

---

Q23. **You detected a large amount of energy emitted by the nucleus of an active galaxy. The energy is emitted by _____.**

a. a supernova

b. the accretion disk just inside the event horizon of a black hole

c. the accretion disk just outside the event horizon of a black hole

d. a cluster of NS

e. dark matter

Q24. **Which of the following is true?**

a. A white dwarf was discovered with a mass of about 10^6 solar masses

b. A supernova explosion left behind a NS of about 10^6 solar masses

c. The center of our galaxy might have a black hole of about 4×10^6 solar masses

d. All of the above are true

Q25. **GRB involve a NS with a companion normal or red giant star.**

a. True

b. False

Q26. **Most of the millisecond pulsars known are inside _____.**

a. globular clusters

b. binary systems

c. planetary nebulae

d. a and b

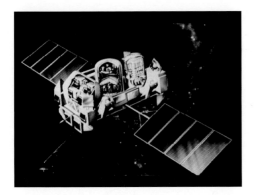

FIGURE 9-12. *The Compton Gamma-Ray Observatory was in orbit from 1991 to 2002.*

Credit: NASA Johnson Space Center Collection

reaches million of degrees, and emit X-rays and other types of radiation before it crosses the event horizon to go into oblivion inside the black hole. Notice that the radiation does not come from the black hole. It comes from the accretion disk around the black hole. This is another type of X-ray binary system, similar to the one created by NS in binary systems.

X-ray binary systems usually are formed by a visible normal star or a red giant and an invisible object at the system's center. The invisible center could be a NS or a black hole. How do we know which one it is if we cannot see it?

The mass of the invisible object can be estimated from the orbit and the orbital motion of the companion star around the invisible center. If the estimated mass of the object at the center ranges from 1.4 to 3 solar masses, it is a NS, and if it is larger than three solar masses, it is a black hole. The same method was applied to determine the mass at galactic centers. The mass at the center has to be large enough to keep the matter in orbit.

The brightest X-ray sources in our galaxy are X-ray binaries. The Cygnus X-1 is an X-ray binary system that might have a black hole at its center.

As we have seen, we are able to estimate the value of the mass of black holes, but we do not yet have any means to determine how matter behaves inside the event horizon of a black hole. This is because the laws of physics break down inside the event horizon. We only know that in a black hole the mass is concentrated in a single point, or singularity, of infinite density.

Q27. Millisecond pulsars _____.
a. are pulsating red dwarfs
b. are fast spinning white dwarfs
c. are produced by molecular clouds
d. involve NS spinning at more than 330 times per second
e. b and d

Q28. What is the difference between gravitational redshift and Doppler redshift?

Q29. Discuss the following statements and establish their truth or untruth.
a. The radii of white dwarfs gradually increase as they accrete matter from the companion
b. If a black hole 20 times as massive as the Sun were lurking near the dwarf planet Iris, we would have no way of knowing it was there
c. We can detect the X-rays emitted by matter falling into a black hole as soon as it crosses the event horizon

ANSWERS

1. b
2. b
3. b
4. c
5. d
6. d
7. a
8. e, P = 1/f = 1/25

9. e, f = 1/P = 1/0.8
10. a
11. c
12. b
13. d
14. e
15. b
16. c
17. c
18. b
19. a
20. c
21. c
22. c
23. c
24. c
25. b
26. d
27. d

UNIT 4

The astronomers of the beginning of the 19th century, were debating whether or not our galaxy was the only galaxy in the universe. It came as a surprise to many of them when Hubble, in 1926, discovered that our galaxy is one of the many that populated the universe.

In this unit we are going study the main properties of the Milky Way, and other galxies.

We are going to learned that the universe has an origin and that it is expanding.

The new generation of telescopes of the last 30 years has extended our view of the universe and have given us detail maps about the distribution of the galaxy trough out the universe.

CHAPTER 10

The Milky Way Galaxy and Other Galaxies

Radio, infrared, and optical telescopes have revealed many of the secrets of the Milky Way and other galaxies. The Hubble Space Telescope has a special place in the exploration of the heavens. In this chapter, we will discover that the heavens are populated with billions of galaxies and that they are intertwined in clusters and superclusters. We will also discover that the spectra of the majority of the galaxies is redshifted and that the universe is continuously expanding. We will begin our study by looking with some detail at the Milky Way.

FIGURE 10-1. *Whirlpool galaxy, M51, NGC 5194/5.*

Credit: NASA Chandra Telescope Collection

THE DIMENSIONS OF THE MILKY WAY

The astronomers of the 19th and the early 20th centuries did not advance much in the knowledge of the distribution of the stars and shape of the galaxy. In general, it was believed that the majority of the stars were located along the galactic disk with a few stars and globular clusters distributed on either side of the disk with the Sun near the center of the disk. The dimensions of the star system, as many referred to the galaxy, were not well-known.

Harlow Shapley, working with the most advanced telescope of his time, the Mount Wilson telescope, estimated the dimension of the Milky Way by measuring the distances to the globular clusters. To measure these distances, he used the variable stars, RR-Lyrae, located in the globular clusters.

To analyze his results in 1917, Shapley mapped the distances that he obtained to the globular clusters and the directions in which they were located. He discovered that the diameter of the galaxy was about 100,000 pc and that the globular clusters were almost uniformly distributed not around the Sun but in the direction of the constellation Sagittarius. This result was unexpected because he thought, as every one else did at the time, that the Sun was close to the galactic center. See Figure 10-2.

> **Q1. To determine the dimensions of the Milky Way, Shapley used _____.**
>
> a. open clusters
> b. molecular clouds
> c. globular clusters
> d. black holes
> e. c and d
>
> **Q2. Which of the following is true?**
>
> a. Copernicus pointed out that the Earth is not at the center of the solar system.
> b. Shapley pointed out that the solar system is not at the center of our galaxy.
> c. Our galaxy is not at the center of the universe. In fact, there is no evidence to indicate that the universe even has a center.
> d. All the above are true.

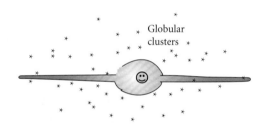

FIGURE 10-2. *General view of the galaxy at the beginning of the 20th century.*

Shapley overestimated the dimensions of the galaxy by about two orders of magnitude because he was unaware of the existence of interstellar dust. Therefore, he did not include the interstellar extinction in the distance calculations. He also thought the variable stars he was using were the more luminous yellow Cepheids variables instead of the less luminous RR-Lyrae. He, however, understood the main points correctly; i.e., the Sun is not the center of the galaxy and the galaxy was bigger than previously thought. Modern measurements locate the galactic center about 8.5 kpc (approximately 25,000 ly) from the Sun and the diameter of the galaxy between 75,000 ly years (23 kpc) and 100,000 ly (30 kpc).

A common belief at the beginning of the 20th century was that the Milky Way was the universe. Therefore, Shapley was convinced that he had determined the dimensions of the entire universe. Soon after Shapley completed his work on measuring the galactic dimensions, another experimental astronomer, Herber Curtis (working also at Mount Wilson observatory), concluded that our galaxy was one of the many galaxies populating the universe.

On April 26, 1920, Shapley and Curtis participated in what is called the **Great Debate** of astronomy. Curtis was in favor of a universe formed by many galaxies, and Shapley was in favor of a smaller universe consisting of only one galaxy. The records do not include details about the winner of the debate. However, a few years after the debate, in 1923, Edwin Hubble working at Mount Wilson observatory demonstrated beyond any doubt that Curtis was right. Shapley measured the distances to the variable stars in the globular clusters using the equation that gives the distance to a star, knowing its apparent and absolute magnitude (See equation 6-1). This equation was explained and illustrated in chapter eight. The value of apparent magnitude of the normal and variable stars was known and available to Shapley, but he also needed the absolute magnitudes, which were not well established at that time. Shapley was able to construct a period-luminosity relation to obtain the absolute magnitude of the variable stars. The calibration curve obtained by Shapley is similar to the period-luminosity diagrams shown in Figures 8-19 and 8-20 in chapter eight. Problems 10-3 and 10-4 give a review of the method used by Shapley to measure the distance to variable stars, using the well-known 6-4 equation.

$$d = 10^{\frac{m-M+5}{5}} \text{ (pc)}.$$

Structure of the Milky Way

The Milky Way a has a halo, a central bulge, and a disk as shown in Figure 10-4. In addition to the visible parts, it also has an invisible component called **dark matter**. The stars in the galaxy fall into two types of populations. Population I is second generation of stars and thus has a relatively large concentration of elements heavier than helium. Astronomers sometimes call the elements heavier than helium heavy metals. So Population I stars are **metal rich**. The first-generation stars are **metal poor** and are called **Population II stars**. The transition from one group to the other is so gradual it creates an intermediate group of stars in which there is not a clear cut difference between the two groups.

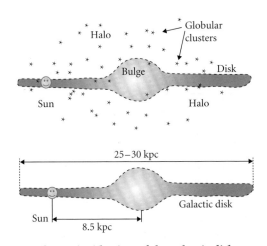

FIGURE 10-4. *Schematic side view of the galactic disk.*

Q3. A Cepheid variable with a period of 20 days, Figure 10-3, has an absolute magnitude of about _____.

a. −4
b. +4
c. −5
d. +5
e. −6.0

Q4. How far is a globular cluster that contains a Cepheid variable whose luminosity changes with a period of 70 days, and has an apparent magnitude of 3.5? Use Figure 10-3.

a. 1,000 pc
b. 1,000 ly
c. 10,000 pc
d. 10,000 ly
e. 20,000 pc

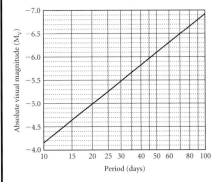

FIGURE 10-3. *Period-absolute magnitude relationship for Cepheid variables.*

Chapter 10—*The Milky Way Galaxy and Other Galaxies*

The Disk and the Spiral Arms of the Galaxy

The galactic disk contains most of the gas and dust of the galaxy. It also contains the molecular clouds where stars form, the open clusters, stellar associations, and the young bright main sequence stars. The best estimates for the diameter of the galactic disk puts it between 25 and 30 kpc (approximately 75,000–100,000 ly). The thickness is not uniform—near the Sun the thickness is only about 350 pc, and it grows thicker towards the center. The Sun is about 8,500 pc from the galactic center.

The galactic disk contains the spiral arms, where the molecular clouds are located and where the stars are forming.

The star formation in the disk started about 8 billion years ago, and still it is going on. The stars in the disk are second- or third-generation stars and most are Population I stars. However, it also has a few old Population II stars.

How do we know the *galaxy* has arms? The spiral traces reveal the spiral structure of the Milky Way. There are optical and radio spiral tracers. The first group includes molecular clouds, HII regions, open clusters, bright O and B stars, and the second 21-cm radio radiation. The galactic dust prevents us from observing the optical traces beyond a few tens of parsecs away from the Sun. Optical tracers can only be observed in the Sun's neighborhood. Telescopic observations close to the Sun indicate that these spiral traces are distributed along arcs, as shown in Figure 10-5. How do we know that these features extend over the entire galactic disk?

Dust is transparent to radio and infrared wavelengths, so the entire galactic disk can be studied in these two wavelengths.

The maps obtained since the 1950s using the 21-cm radio radiation indicate that the atomic hydrogen in the galaxy is distributed along four major arms, called **Norma**, Scutum-Centaurus, Sagittarius, the Orion or local arm and Perseus. (Recall that 21-cm radio radiation is related to the spin flip of cold atomic hydrogen.)

Radio observations of carbon monoxide molecules inside the molecular clouds also show that the molecular clouds are distributed along galactic arms, confirming the spiral structure of the galaxy.

Recent observations since the 1990s in the infrared done with the NASA's Spitzer Space Telescope indicate that the Milky Way

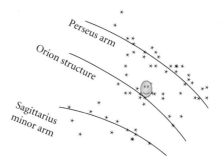

FIGURE 10-5. *Near the Sun, the optical spiral tracers outline the presence of the spiral arms.*

has only two major arms and two minor less-populated arms. The two major arms are the Scutum-Centaurus and Perseus and have the largest concentration of stars. The other two arms Sagittarius and Norma are two minor arms, or filaments, that are not part of the galaxy's dominant spiral structure and have lower concentration of stars. The two larger arms, Scutum-Centaurus and Perseus, originate at the end of a central bar of stars.

Our Sun seems to be located on the edge of a structure called the **Orion Spur** of the minor Sagittarius arm, between the Perseus arm and the Scutum-Centaurus arm.

The spiral structure is a common feature in many other galaxies. Notice that the arms of this galaxy are very bright. The brightness is due to the presence of young stars and open clusters, throughout the disk.

The Galactic Bulge

The galactic disk is about 350 pc thick near the Sun, but the thickness increases toward the center forming the bulge. See Figure 10-4. The dust in the galactic disk makes it difficult to observe the bulge at optical wavelengths. See Figure 10-4. The Cosmic Back Ground Explorer (COBE) observed the entire Milky Way in the infrared and revealed that the bulge has an elongated shape. See Figure 10-6. The bulge contains Population I and Population II stars of different ages. The youngest group has ages less than 200 million years the intermediate group has ages between 200 million years and 7 billion years, and the third group are stars older than 10 billion years. The youngest stars are toward the center, and the oldest are in the outer regions where some globular clusters have been observed. Observations indicate that the number of stars increases toward the center of the bulge. The bulge is very luminous because it has many young stars. The bulge contains a bar of stars, of about 9 kpc long, from whence the two major galactic arms emerge.

FIGURE 10-6. *The COBE satellite infrared view of the Milky Way galaxy.*

Credit: The COBE Project/DRIBE/NASA

The Halo

The halo is a semispherical region of about 75,000–100,000 ly in diameter surrounding the entire visible part of the galaxy. It also contains individual stars as well as the globular cluster. Astronomers have discovered 150 globular clusters to present. See Figure 10-4.

Q5. If you want to study the galactic bulge, you will use a (an) _____ telescope.

a. optical
b. X-ray
c. infrared
d. radio
e. c and d

Q6. Which of the following is not used as a spiral tracer?

a. white dwarfs
b. atomic hydrogen
c. open clusters
d. molecular clouds
e. c and d

Q7. The disk and arms of spiral galaxies are very bright and luminous because _____.

a. they are rich in oxygen
b. they are sites of star formation
c. they are home to the open cluster
d. of the numerous supernovae that are occurring there
e. b and c

Q8. The arms of the Milky Way seem to originate from a central bar.

a. True
b. False

TABLE 10-1. *Comparing the characteristics of the two stellar populations.*

	POPULATION I	POPULATION II
Orbits	Almost circular in disk	Very elliptical, in the halo
Distribution	Mainly in open clusters	Mainly in globular clusters
Location	Spiral arms	Halo, bulge
Composition of elements heavier than helium	Rich in heavy metals, betweeen 3 and 4%	Poor in heavy metals less than 0.4%
Age (in years)	Very young, less than 7 billion years old	Atleast 10–12 billion years old
Typical object	Molecular clouds, open clusters, HII regions, O and B stars	Globular clusters, RR-Lyrae star

The halo is almost gas and dust-free, and it does not contain molecular clouds. Star formation in the halo ceased more than 10 billion years ago, when most of the gas and dust were exhausted. For this reason, it is the least luminous component of the galaxy. The halo stars were the first stars to form in the galaxy and are Population II stars because they are poor in metals. The halo stars are at least 10 billion years old. The properties of the two star populations are summarized in Table 10-1.

The Galactic Nucleus

The galactic nucleus is inside the galactic bulge in the direction of the constellation Sagittarius. The dust in the plane of the galaxy prevents us from observing the galactic center in the visible, but it can be observed in longer wavelengths, such as infrared, microwave, and radio wavelengths. The galactic center can also be studied in the shorter X-ray and gamma ray wavelengths.

The galactic center has young hot stars near the center. The ultraviolet radiation from the hot stars near the center causes the dust to brightly shine in the infrared. The inner region of the galaxy is rich in dust and stars. In general, the galactic nucleus is rich in stars. The number of stars increases toward the center. The center is so crowded with stars that occasionally the stars collide with each other.

Radio observations indicate that the center of the galaxy is the source of several radio sources. The radiation emitted by Sagittarius A* (Sgr A*) is particularly intense.

Sgr A* is a region crowded with stars. The galactic center is also the source of a continuous X-ray emission and of X-ray bursts lasting only about 1 h. Each X-ray burst doubles the X-ray luminosity of the center.

The Sgr A* region is very dynamic and has many stars and abundant gas, and dust orbiting close to it. Studying the orbital motion of the so-called **S2 stars** around Sgr A*, astronomers have found that Sgr A* has a mass approximately 4 million solar masses.

Several alternatives have been proposed to explain the origin of the mass and of the relative large amount of energy that originates at the galactic center. These alternative proposals include clusters of stars, clusters of neutron stars or white dwarfs, and even the presence of dark matter. None of these proposals have been able to provide a satisfactory explanation. The only alternative left is to assume that the mass is in the form of a supermassive black hole of about 4 million solar masses, with an event horizon of 12 million km. The existence of a black hole provides a reasonable explanation for the observed energy from the central galactic region. The energy arises from the black hole's accretion disk. The accretion disk is fed by matter that falls into it from the collision of stars close to the event horizon. Yes, as was mentioned above, the galactic center is crowded with stars. The stars located here, on average, are separated only 0.02 ly, so star collisions are likely to occur. By comparison, the stars near the Sun are separated by approximately 5 ly. The black hole in our galaxy is relatively quiet when compared with black holes at the center of active galaxies.

Motion of the Stars in the Galaxy

The stars in the disk rotate in the same direction, but their speed depends on the distance from the galactic center; i.e., the disk stars have differential rotation. The Sun circles the galaxy in about 240 million years at a distance of 8.5 kpc from the center, with a speed of 240 km/s. The globular cluster in the halo and the few isolated stars it contains have random motions and move with different speeds and different directions in highly elliptical orbits. See Figure 10-7.

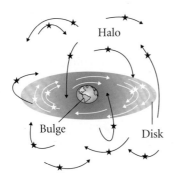

FIGURE 10-7. *Motion of stars in the Milky Way.*

Age of the Milky Way

The globular clusters are the oldest objects found in the galaxy. Therefore, their age gives the age of the Milky Way. Analyzing different globular clusters, astronomers have concluded that the

Q9. Since most of the known globular clusters are in the halo and in the bulge, we can affirm that the halo and the bulge are older than the galactic disk.

a. True
b. False

Q10. There is experimental evidence that the center of the Milky Way is the home of a _____.

a. star of a about a million solar masses
b. supermassive black hole
c. cluster of massive neutron star
d. large cluster of white dwarfs

Q11. The center of an elliptical galaxy has a supermassive black hole of 8.5 million solar masses. The radius of the event horizon of the black hole is _____ million km.

a. 4.8
b. 8.5
c. 19.4
d. 25.5

Q12. The fuzzy band across the sky that we called the Milky Way is actually the light from millions of stars in the _____ of the galaxy.

   a. halo
   b. nuclear bulge
   c. spiral arms and disk
   d. center

Q13. Why does interstellar extinction prevent us from seeing optical tracers beyond a few 1,000 pc?

   a. The galaxy is smaller than that distance.
   b. The light is blocked by the interstellar matter.
   c. Optical tracers are faint.
   d. The Sun is in the galactic halo.
   e. Optical telescopes cannot see beyond this distance.

TABLE 10-2. *Comparing the main properties of the galactic disk, halo, and bulge.*

GALACTIC DISK	GALACTIC HALO	GALACTIC BULGE
Highly flattened	Roughly spherical—mildly flattened	Somewhat flattened and elongated in the plane disk of the ("football shaped")
Contains both young and old stars and the open clusters	Contains the globular and old stars only	Contains both young and old stars. Younger stars are toward the inner regions
Contains most of the dust and gas of the galaxy	Almost free of gas and dust	Contains gas and dust, especially in the inner regions
Sites of ongoing star formation	No star formation during the last 10 billion years	Ongoing star formation in the inner regions
Gas and stars move incircular orbits in the galactic plane	Stars have random orbits in three dimensions	Stars have largely random orbits, but with some net rotation about the galactic center
Spiral arms	No obvious substructure	Ring of gas and dust near center; contains the galactic nucleus
Overall white coloration, with blue spiral arms	Reddish in color	Yellow-white

galaxy is about 13 billion years old. Recall that the age of a cluster is obtained from the star located at the turn-off point, on the HR diagram. This was discussed in chapter seven. The open clusters and star associations are much younger than the globular cluster. They are less than 7 billion years old and cannot be used to determine the age of the galaxy. Table 10-2 gives a summary of the main properties of the galaxy.

The Mass of the Galaxy and Dark Matter

To determine the mass of the galaxy, we assume that most of the visible mass of the galaxy is in the galactic disk and inside the orbit of the Sun. This is not a bad approximation because the halo's mass is very small compared with the mass in the disk, and the Sun is toward the edge of the galaxy. See Figure 10-8.

Elementary physics dictates that the motion of the Sun around the galactic center is controlled by the mass of the galaxy inside the

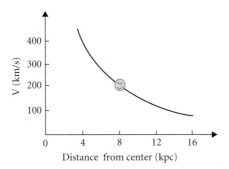

FIGURE 10-8. *Orbital speed of the stars if they had Keplerian orbital motion.*

Sun's orbit. The mass outside the Sun's orbit does not affect the motion of the Sun. Therefore, we can assume that the entire mass of the galaxy is concentrated in the galactic center. Under these conditions, the mass of the galaxy can be found using the modified Kepler's third law:

$$\text{Mass} = \frac{(\text{distance from Sun to galactic center})^3}{(\text{period of Sun around galactic center})^2}$$

The distance is in AU, and the period is in years. The Sun orbits the galactic center in 240 million years at a distance of 8.5 kpc (which equals approximately 1.75 billion AU). Replacing these values in the previous equations, we find that the visible mass of the galaxy in the disk and inside the orbit of the Sun is about 100 billion solar masses. The total visible mass of the galaxy is a little bit larger than this value because we have neglected the stars and dust outside the orbit of the Sun and the mass of the stars in the halo.

The 100 billion solar masses are the lower limit to the galactic mass of the visible mass of the galaxy. Approximately 10% of this mass is in the form of interstellar dust, gas, and molecular clouds. The remainder mass is in the form of stars. If all the masses of the galaxy were distributed inside the visible edge, then the orbital speed of the stars and gas would decrease with distance from the center. This happens with the speed of the planets in the solar system. This type of rotation is known as **Keplerian motion**. In this case, the rotational speed of the stars around the center of the galaxy would be described by the diagram in Figure 10-8. Observations show otherwise. When the speed of the rotation of the stars in the Milky Way is measured and plotted as a function of the galactic distance from the center, the speed of rotation of the stars beyond the Sun does not go down but remains flat and even increases a little bit with distance. This is illustrated in the rotation curve of Figure 10-9. This tells us that the mass of our galaxy is larger than the luminous or visible mass, indicating that beyond the traditional galactic visible edge there is mass that we cannot see. This invisible mass is called **dark matter** and is mainly distributed over the galactic halo and beyond.

The dark matter is invisible because it does not emit any electromagnetic radiation. We know that it exists because its gravitational

Q14. The rotation curve of the galaxy indicates _____.

a. that the speed of rotation of the stars decreases with distance from the galactic center

b. that the speed of rotation of the stars levels off and slightly increases with distance from the galactic center

c. that the galaxy has unseen matter that holds the stars in orbit

d. b and c

Q15. If most of the mass of the galaxy were inside the orbit of the Sun, then we would expect the orbital speed of the stars to decline at greater distances, as shown in Figure 10-8.

a. True

b. False

Q16. Most of the dark matter in the galaxy lies in the disk and galactic center.

a. True

b. False

Q17. Dark matter has not been observed at any electromagnetic wavelength.

a. True

b. False

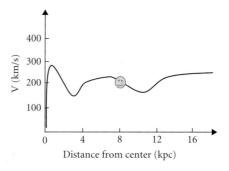

FIGURE 10-9. *The Milky Way's rotation curve. Notice that the speed of the stars beyond the Sun increases with distance.*

force keeps the stars in orbit around the center of the galaxy. If the dark matter were not present, the stars at the edge of the galaxy will fly apart with time which would destroy the galaxy. Obviously, this is not happening. Most of the dark matter is beyond the traditional visible edge of the galaxy in an extended halo or galactic corona. The radius of the galactic corona is quite significant. It might extend about seven times the visible radius of the galaxy and could contain up to 2 trillion solar masses. We do not know the nature of dark matter. It has been suggested that the dark mass in the galactic corona consists of faint old low mass red stars, or Jupiter-sized objects, that are difficult to detect because they are small and very far away. Other theories propose that dark matter is made of some exotic subatomic particles called **weakly interacting massive particles** (WIMP). In summary, the visible mass of the *galaxy* is at least 100 billion solar masses, and the total mass (including normal and dark) is about 600 billion solar masses. A significant amount of dark matter exists. Other galaxies studied so far contain dark matter, and about 90% of the mass of the universe is dark matter. So you, me, the stars, and everything we see are part of the 10% visible matter of the universe! We know that dark matter exists because of its gravitational effects on the stars.

ORIGIN OF THE MILKY WAY

Some astronomers believe that the Milky Way formed from a huge protogalactic cloud of gas, mainly hydrogen and helium that collapsed under its gravitational force. This protogalactic cloud was several 10 kpc across and had a temperature of only a few Kelvin. As the cloud was collapsing, the globular cluster and the halo stars formed. These were the first generation of stars, or Population II. Many of the stars that formed in the halo were high-mass stars, which go quickly through their life cycles, and some end up as supernovae. As these stars died, they enriched the interstellar mater with heavier elements than helium. The gas that did not form halo stars kept falling and formed a central bulge and a disk. The stars that formed later on in the disk incorporated in their bulk the heavier elements formed by the death of the previous generation of stars. The disk stars are mainly Population I stars and have ages that range from a few million to about 10 billion years old. The halo stars are older than 10 billion years. It is important to point out that the disk also contains some old Population II stars, and that the bulge also contains some young Population II stars. This model clearly explains the difference between Population I and Population II stars. However, the classification of the stars into two different populations is an over simplification because there is a continuous variation in stellar ages from the old halo stars to the youngest stars inside molecular cloud. Other models about the formation of the galaxy include the merger of smaller protogalactic clouds, where globular clusters were already forming in the moment of the merger.

OTHER GALAXIES

As already discussed, some astronomers, at the beginning of the 20th century, thought that the universe consisted of a large star system whose center was the solar system. Harlow Shapley discovered in 1917 that the Sun was not the center of the galaxy but that it was located on the disk at about 8.5 kpc away from the center. Even after this remarkable discovery, Shapley erroneously believed that the galaxy was the universe. The telescopes that he used did not have enough resolution to distinguish that some of the objects that he called nebulae were actually other galaxies. In 1923, Edwin Hubble discovered that our galaxy is one of the many star systems that populate the universe. The universe is vast and contains an unknown number of galaxies separated by incredible huge distances. Everywhere we point our telescopes we see several galaxies.

Classification of the Galaxies

In 1929, Edwin Hubble classified the galaxies according to their appearance: **spirals** (**normal** and **barred**), **elliptical**, and **irregulars**. Hubble arranged the different types of stars in a tuning fork. See Figure 10-10. (You should notice that Hobble's tuning fork

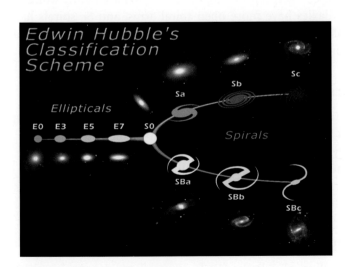

FIGURE 10-10. *The Hubble's fork.*

Credit: NASA Hubble Space Telescope Collection

Q18. Up until now, astronomers do not have any experimental evidence that dark matter mass exists.
  a. True
  b. False

Q19. Most of the mass of the Milky Way seems to exist in the form of _____.
  a. Population I stars in the disk
  b. Population II stars in the halo
  c. hydrogen gas in the disk and spiral arms
  d. dark matter out in the dark halo
  e. the black hole in the galactic center

Q20. What is in the center of our galaxy and why do we think this?
  a. A star forming region is in the center of the galaxy, and we think this because we can see the Herbig-Haro objects created by the newly formed stars' protostellar jets.
  b. A very massive black hole is in the center of the galaxy, and we think this because we can calculate the object's mass from the orbital motions of stars near to it.
  c. An A-type main sequence star is in the center of the galaxy, and we think this because we can see the hydrogen absorption lines in its spectrum.
  d. A neutron star is in the center of the galaxy, and we think this because we can see its pulses.

does not represent any sequence of evolution between the different galaxies.)

Spiral Galaxies

Spiral galaxies contain a disk, arms, and probably a halo. The halo of the galaxies, except in the case of the Milky Way, cannot be observed because the distance to the galaxies is of the order of million of light years. The disks of the spiral galaxies contain large volumes of gas and dust, as seen in the edge on the image shown in Figure 10-11. The disks, arms, and bulge of the spirals are bright due to the presence of the hot, young O and B stars. Most spirals are smaller than the Milky Way, and only a few are larger and more luminous. The spiral galaxies make about two-thirds of all bright galaxies. The spiral galaxies are divided into two main classes: 1) the normal spirals (S) and 2) the barred spirals (SB). Both of these are subdivided into three different subclasses: a, b, and c. This classification is based on the following criteria:

1. the tightness of the spiral arms and
2. the relative size of the central bulge.

FIGURE 10-11. *The edge-on view of the spiral galaxy NGC4013 shows the presence of much dust along the disk.*

Credit: NASA Hubble Space Telescope Collection

Normal Spirals

Normal spirals come in three subclasses: Sa, Sb, and Sc. Sa galaxies have the largest central bulge and tend to have tightly wound arms. Sb galaxies have more open spiral arms, and Sc spirals have the smallest central bulge and loose open spiral arms. The Sb galaxies are intermediate between Sa and Sc.

Barred Spirals

One of the most common type of spiral galaxies is the **barred spiral**. In a barred spiral, the arms spring up at the end of a bar of stars. They also come in three different types: SBa, SBb, and SBc. The S means spiral, the B means barred spiral, and the third letter refers to the relative size of the central bulge and the degree of tightening of the arms.

Elliptical Galaxies

The elliptical galaxies (E) have an ellipsoidal shape, which is a kind of a deformed sphere. Some are very elongated and others are almost round, as shown in Figure 10-12. These galaxies do not show obvious structure and have neither arms nor disk. They have small amounts of cold interstellar matter, dust, and gas. They do not show evidence of ongoing star formation. Most of the stars in these galaxies are cool, red, and old. The stars in these galaxies have three-dimensional random motion, which is similar to the motion of the stars and globular cluster in the halo of the Milky Way. Elliptical galaxies exist in a wide range of sizes and masses, from dwarfs to giant elliptical. The dwarf elliptical galaxies exceed the larger ones. They are only about one-tenth the diameter of the Milky Way and contain fewer than a million stars. On the other hand, the giant elliptical galaxies have diameters of a few megaparsecs and contain trillions of stars. The largest known elliptical galaxy is the M87, or NGC-4486, as shown in Figure 10-12. It is believed that the jet going out from the surface of the galaxy is produced by a supermassive black hole of 2 billion solar masses at the center of the galaxy. Elliptical galaxies have zero to low rotations, while the spirals have considerable amounts of rotation. The greatest variation in size and mass occurs in elliptical galaxies.

FIGURE 10-12. *The elliptical galaxy M87 is near a sphere, while.*

Credit: NASA Hubble Space Telescope Collection

Irregular Galaxies

The galaxies that have irregular shapes have neither arms nor bulge, and the stars within the galaxy are randomly distributed without any obvious order. About 50% of the mass of the irregular galaxies is in the form of interstellar matter, gas, and dust. Therefore, star formation is very active. The irregular galaxies have both old and young stars. They are highly luminous because they have several young hot stars. Star formation is very active in this galaxy. The galaxy sparkles with the light from millions of newly formed young stars. The Large Magellanic Cloud prominent in the southern sky and only 51 kpc away, is a good example of an irregular galaxy. It was in this galaxy that the 1987 supernova explosion discussed in chapter nine was observed. Most of the irregular galaxies are smaller than the spiral galaxies, and they have masses between 1,100 million and 10 billion solar masses.

TABLE 10-3. *Comparison of the three different types of galaxies.*

	SPIRAL/BAR SPIRAL S/SB	ELLIPTICAL (E)	IRREGULARS (Ir)
Structure Gas and dust	Disk, arms, and halo. Disk contains gas and dust. Halo dust, gas free.	No disk or arms. Spheroid shape. Small amount of gas and dust.	Irregular structure. Lots of gas and dust.
Star formation	Stars form in disk. young and old stars.	No star formation. Old stars.	Active star formation. Mainly young stars.
Stellar motion	In disk stars and dust move in circular orbits. Halo stars have radon motions.	Stars have radon motions in three dimensions	Stars move in different dimensions
Sizes	Most smaller than the Milky Way	Giants, large, and small	Smaller than the spiral galaxies

Note the following systematic trends:

1. Interstellar material ranges from essentially no material in ellipticals to more material in spirals and substantial quantities of material in some irregulars.

2. Ellipticals have low to zero rotation rates, while spirals have relatively high rotation rates.

3. The most massive galaxies are ellipticals. However, there are many small dwarf ellipticals.

Irregular galaxies are smaller than the Milky Way and have less number of stars. Spirals tend to be intermediate in size—between ellipticals and irregulars.

DISTANCE DETERMINATION TO GALAXIES

The distance to nearby galaxies can be found using the yellow Cepheid variables. This method, which was explained earlier, can be used to measure distances up to 25 Mpc, which correspond to distances within the local supercluster. For larger distances, other methods are used. In what follows, you will see the description of only two different methods. The first method uses supernovae Type Ia, and the second method uses the redshifts of galaxies.

Supernovae Type Ia and Distance Measurements

We saw earlier that supernovae Type Ia are always produced by white dwarfs as they approach the 1.4 solar mass limit, and that their peak light output, always corresponds to an absolute magnitude −19.6.

> **Q21. A supernova Type Ia of apparent magnitude 10.4 was observed in a galaxy. How far away is the galaxy?**
>
> a. 1,000 pc
> b. 1,000 ly
> c. 10,000 pc
> d. 10,000 ly
> e. 100,000 pc

Since supernova Type Ia are very luminous they can be observed in distant galaxies, and are used to measure the distance to this galaxies.

When we observe a supernovae Typa Ia, we know that they have an absolute magnitude of -19.6, so if we measure its apparent magnitude at the maximum light output we can find its distance using the well-known equation that relates distance, apparent magnitude and absolute magnitude.

$$d = 10^{\frac{m-M+5}{5}} \text{ (pc)}$$

Supernovae Type Ia are about 14 magnitudes brighter than Cepheid variables and can measure distances to around 1,000 Mpc. This is a significant fraction of the radius of the known universe. See Problem 21.

The Hubble Method of Measuring Distance to Galaxies

In 1929, Edwin Hubble, working with the Mount Wilson telescope, discovered that the absorption lines of distant galaxies were redshifted; i.e., they were receding. He also found that the speed of recession of the galaxies is proportional to their distance, meaning that the faster a galaxy moves, the more redshift it shows and the farther the galaxy is. It is important to realize that the speed of the galaxies is not directly observed with a telescope. We measure their redshift and then determine the speed of recession using the Doppler equation. (Review the Doppler effect in chapter three.) In order to find a relationship between the speed and the distance, Hubble built a calibration curve from experimental observations. Hubble discovered that the distance to a galaxy is related to the speed of recession by the following simple law:

$$V = H_0 D,$$

where **V** is the recession speed, **D** is the distance to the galaxies in megaparsecs, and H_0 is the Hubble's constant. This relation is known as the **Hubble's law**.

The value of the Hubble's constant is controversial. The current accepted values goes from 50 to 75 km/s/Mpc. Many astronomers claim that the best value for **H** is 71 km/s/Mpc. A galaxy 1 Mpc away will be moving away from us at a speed of 71 km/sec, while another galaxy 20 Mpc away will be receding with a speed of 1,420 km/s.

Example

The redshift of a galaxy indicates that it is receding with a speed of 15,000 km/s. If **H** = 71 km/s/Mpc, find the distance to the galaxy.

By solving the previous equation for **D** and replacing the values, we derive

$$D = V/H$$
$$= [15{,}000 \text{ km/s}] / [71 \text{ km/s/Mpc}]$$
$$= 211 \text{ Mpc}.$$

The galaxy is 211 Mpc away.

The redshift of galaxies, along with the Hubble law, has been used to make a large survey of the positions of the galaxies in the universe. See the next section.

THE EXPANSION OF THE UNIVERSE

Hubble's law is a well-established experimental law. There are two important consequences of this law:

One: The universe is expanding. This conclusion is evident because the galaxies are moving away from us, except for a few in the local cluster. This does not mean that we are the center of the universe. If you were in a remote galaxy, you would see the same phenomena. Every observer in the universe sees the clusters of galaxies moving away with a speed proportional to the distance. The galaxies appear to be moving away because the space itself is expanding and carrying the galaxies. It is like raisins in a fruit cake moving away from each other as the cake bakes and grows. From the view point of any raisin, the others are receding. The farther away they are from each other, the faster they move. See Figure 10-13.

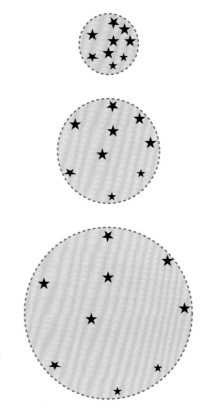

FIGURE 10-13. *As space expands, the galaxies move away from each other. The rate expansion of space corresponds to the rate at which the galaxies move.*

Two: Since everything is moving away from everything else, we conclude that in a distant past, everything was closer together. Yes, there was a time when everything in the universe was concentrated in a single point or singularity. Then the universe begun suddenly with a **Big Bang**. When was all the matter of the universe concentrated in a single point? To find the answer, we assume that the universe has been expanding at a constant rate since the Big Bang, and that the value of the Hubble's constant has not changed since the creation of the universe. This is a daring assumption because we do not know if that has been the case. In any event, let's call $t = 0$ when the Big Bang happened. Then, in a time t, the universe has expanded a distance

$$D = Vt$$

Using Hubble law to replace V ($V = H_0 D$), we get

$$V = H_0 (Vt)$$ and canceling V we find that

$$1 = H_0 t$$

and finally solving for t, we obtain that the age of the universe is given by

$$t = \frac{1}{H_0} \quad (10\text{-}1)$$

The age of the universe depends on the value of the Hubble's constant. We will use for H the value of 71 km/s/Mpc. We need to do a few conversions of units. Here are the steps:

Express Mpc in km: $Mpc = 10^6 pc = 3.085 \times 10^{19}$ km.

$$t = \frac{1}{H_0} = \frac{1}{71\frac{km}{\frac{Mps}{s}}} = \frac{1}{71\frac{km}{3.085 \times 10^{19}\frac{km}{s}}}$$

$$= \frac{3.085 \times 10^{19} \text{ s}}{71} = 4.33 \times 10^{17} \text{ s} = 13.8 \times 10^9 \text{ years}.$$

This value is not exact because the velocity of expansion has changed with time under the influence of gravity, and the value of the Hubble constant also changes with time. For example, astronomers have recently found evidence that the rate of expansion of the universe is increasing.

The Big Bang marked the beginning of the universe. But where did it happen? Before the Big Bang, there was no space and no time. They were created during the Big Bang. Therefore, the Big Bang happened everywhere and at once. The expansion of the universe began soon after the Big Bang at the end of the so-called **inflation phase**.

Clusters of Galaxies

The galaxies are not randomly scattered in the universe. Rather they tend to congregate in groups or clusters, held together by

---

Q22. A galaxy with a recessional speed of 30,000 km/s with respect to Earth is at a distance of about _____. Use $H_0 = 71$ km/s/Mpc.

  a. 100 Mpc
  b. 600 Mpc
  c. 423 Mpc
  d. 526 million ly

Q23. What is the speed of recession of a galaxy that is 240 Mpc away? Use $H_0 = 71$ km/s/Mpc.

  a. 22,000 km/s
  b. 17,040 km/s
  c. 220,000 km/s
  d. 170,040 km/s

Q24. The absorption lines of distant galaxies are _____, indicating that the galaxies are moving away from each other.

  a. blueshifted
  b. redshifted
  c. well defined
  d. broad and weak

Q25. The value of the Hubble constant is used to determine the age of the universe.

  a. True
  b. False

> **Q26. The Coma cluster is an example of _____.**
>
> a. an elliptical galaxy
>
> b. an irregular galaxy
>
> c. a poor cluster
>
> d. a rich cluster
>
> **Q27. The closest galaxy to the Milky Way is _____.**
>
> a. the Andromeda galaxy
>
> b. a large globular cluster
>
> c. the Small Magellanic cloud
>
> d. a giant elliptical galaxy
>
> e. the Canis Major dwarf galaxy

mutual self-gravitation. The individual galaxies move around their common center of mass. The clusters formed about a more 10 billion years ago. The groups of galaxies have less than 50 galaxies. The clusters have larger number of galaxies. The rich clusters have thousands of galaxies. The Virgo cluster is a good example of a rich cluster. The rich clusters usually have an elliptical galaxy at their center. In these clusters, the number of galaxies increases toward the center.

The Local Group

The Milky Way is part of the local group of galaxies. It has about 35 galaxies inside a volume of about 1 Mpc (1 million parsecs) in diameter. The group has three large spirals, the Milky Way, M31 or Andromeda, and M33 or Triangulum. The Andromeda is approximately 50% bigger than the Milky Way. There are also 13 irregular galaxies. The Large and Small Magellanic Clouds are the largest among the irregulars and the most luminous after the three spirals. The Andromeda and the Milky Way galaxies, separated by a distance of 770 kpc, are approaching each other at a velocity of 119 km/s. Anticipated collision time is 6.3 billion years. The closest galaxy to the Milky Way is the dwarf irregular Canis Major galaxy at a distance of 42,000 ly from the galactic center, in the direction of the constellation Canis Major. For many years, it was believed that the Large Magellanic Cloud, at a distance of about 51 kpc (approximately 150,000 ly), was the closest galaxy.

The Virgo Cluster

The Virgo cluster, about 16 Mpc away, is a rich cluster with over 2,000 galaxies lying inside a volume of 3 Mpc in diameter. The cluster has spiral and elliptical galaxies, with most of the galaxies toward the center. The giant elliptical galaxy, M87 (see Figure 10-12), is near the center of the cluster. This galaxy has a diameter of about 770 kpc, (The diameter is as large as the distance between the Andromeda and the Milky Way). The cluster is important because it is the nearest rich cluster to Earth. The Abell 2218 cluster is located about 3 Mpc away and is another example of a rich cluster. The Abell 2218 is so massive that it bends the light and acts as a gravitational lens.

SUPERCLUSTERS OF GALAXIES AND LARGE-SCALE SURVEYS

The clusters of galaxies are organized in larger groups called **superclusters**. These are clustering of clusters. A good example of these large structures is the Virgo supercluster. This is also known as the **local supercluster** because we are a part of it. We are located towards edge of the supercluster. The center of the supercluster is the Virgo cluster. The supercluster contains about 100 clusters of galaxies and is spread out in a region of about 32–60 Mpc

(100–200 million ly) in diameter and approximately 10^{15} (a thousand trillion) solar mass. Another well-known supercluster is the Perseus-Pices supercluster located about 70 Mpc. About 75% of the galaxies are members of clusters. Clustering of galaxies is very common.

Using the redshift of galaxies and the Hubble law, astronomers at the Anglo-Australian Telescope observatory have mapped the distribution of more than 200,000 galaxies in the sky, all within a distance of 2 billion ly. This survey includes the northern and southern hemispheres, and the project is known as the **2dF Galaxy Redshift Survey** (2dFGRS). Instead of surveying the entire sky, a few wedge-shaped slides have been mapped. Each wedge is 6 degrees. Redshift data can be converted to positions in space. Therefore, the survey has created the most comprehensive three-dimensional map of the local universe. The 2dFGRS has produced a three-dimensional map of the distribution of 221,000 galaxies in the sky. You can see the maps of the survey at: http://www2.aao.gov.au/~TDFgg/.

The analysis of the survey indicates that the galaxies and galaxy clusters are located on the surfaces of huge bubbles of almost empty space called **voids**. The voids are three-dimensional regions, and the galaxies are on the wall of these bubbles.

To visualize the three-dimensional survey, think of a froth of soap. The two-dimensional set of points represents what you will see when you cut across the froth. The voids, or bubbles, are almost spherical regions and have a diameter between 50 and 100 Mpc.

The filaments observed in the two-dimensional map are the galaxies on the walls of two adjacent bubbles. Where several voids meet, the density of points increases. These represent the superclusters of galaxies. Large-scale observations seem to dominate the universe. The visible universe extends to a distance of about 13.7 billion ly. The motion of the galaxies in the superclusters indicates the presence of dark matter in between the galaxies of the superclusters. Without the dark matter, the galaxies would move too fast and they would fly apart.

Galaxy Interactions and Collisions

In a rich cluster, where the individual galaxies are relatively close to each other, collisions and mergers are possible. In a few billion years, the two galaxies will merge into one galaxy. During a collision, the individual stars of the galaxies typically do not collide because the distance between individual stars is very large. However, the gravitational interaction rearranges the individual stars shifting their positions within the galaxies. The gas and dust of the two galaxies collide and mix. Often the collision produces a shock wave that propagates through the merging galaxies, triggering the formation of new stars. This can be seen in the Hubble Space Telescope image of the Antennae galaxies. The merging of galaxies takes millions of years, so we are not able to witness an entire collision.

Mass and Dark Matter in Galaxies

The mass of nearby spiral galaxies can be determined from the rotational velocity of the gas in the disk of the galaxies. At large distances from the galactic center of the galaxies, the rotation speed of the gas, the stars, and the dust should drop according to Newton's law. However, direct measurements indicate that the rotation velocity remains constant and even increases a little bit with distance, as shown in Figure 10-14. The rotation curve of the galaxies tells us, as it was the case for the Milky Way, that the mass of the galaxies is much larger than the visible mass. This excess of mass is what we called earlier **dark matter**. Some astronomers believe that dark matter forms a **dark halo**, or **dark galactic corona**, around the galaxies. The mass of the galaxies in a cluster can be estimated by measuring the rotation velocity of the individual galaxies around their common center of mass. The result also indicates that the clusters contain dark matter. If dark matter did not exist, then the stars in the galaxy and the galaxies in a cluster would fly apart. The mass of the galaxies and galaxy clusters can also be estimated from the luminosity of the galaxies, but always the luminous mass is much less than the mass obtained from the rotation curve. The difference is the dark matter. Dark matter, as stated earlier, is not ordinary matter in that it does not emit electromagnetic radiation. The only effect that can be felt is the gravitational force that it exerts on matter. Dark matter is real. Approximately 90% of the mass of the universe is in the form of dark matter.

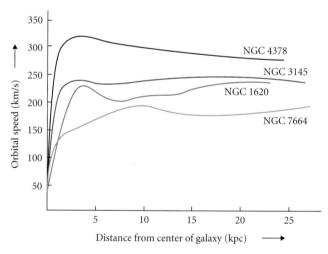

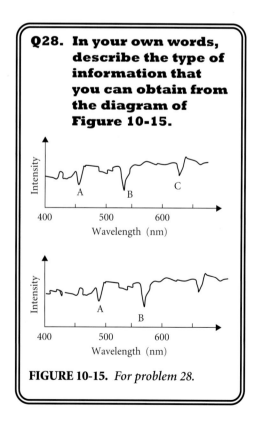

Q28. In your own words, describe the type of information that you can obtain from the diagram of Figure 10-15.

FIGURE 10-15. For problem 28.

FIGURE 10-14. The rotation curve of the galaxies tells us that all the galaxies have dark matter. Neil Comins, Freeman Company 2009.

Look-Back Time

We have seen that the galaxies are very far away. For example, the Cygnus A galaxy is about 750 million ly away. We have seen this galaxy the way it was when the photons left the galaxy, approximately 750 million years ago. The time the light takes to travel from a given object to us is the **look-back time**. The Sun has only 8.3 min of look-back time. The farther out we look into the universe, the farther back in time we see.

Active Galactic Nuclei

The galaxies that we have studied so far are known as **normal galaxies**. There is also a group of elliptical and spiral galaxies that radiate large quantities of energy from a tiny variable nuclei. These galaxies are called **active galactic nuclei** (AGN). For simplicity, we will call them active galaxies.

The luminosity of a normal galaxy comes from the entire body of the galaxy. This is dependent on the individual stars that are spread over the bulk of the galaxy. In the case of active galaxies, most of the radiation is emitted by a tiny variable nuclei. Most of the energy emitted by active galaxies is nonstellar.

There are three types of active galaxies: Seyfert, radio, and quasars. About 10% of the known galaxies are active galaxies.

Seyfert Galaxies

Seyfert galaxies are usually active spiral galaxies. Their enormous luminosity comes mainly from a small variable nuclei. The nuclei of some Seyferts radiate about 10,000 times more energy than the Milky Way Galaxy. The radiation has been observed to double or halve in less than a year. Because the radiation or luminosity of a Seyfert galaxy changes in a short period of time, we conclude that the source of radiation has to be small. We arrive at this conclusion because the time in which the radiation changes cannot be smaller than the time the radiation takes to transverse the diameter of the source that produces it. Objects whose luminosity changes in a short time cannot be very large. The nuclei are very bright and less than 3,000 ly in diameter. Another feature of these galaxies is the broad emission spectra from their nuclei, indicating that the matter around the central core is rapidly moving around it. Most of the Seyfert galaxies are barred spirals. Additionally, most have companion galaxies with which they might be gravitationally interacting. About 1% of all spiral galaxies are thought to exhibit Seyfert properties. Another way to think about this is that perhaps all spiral galaxies are Seyferts for only 1% of their lifetime.

Radio Galaxies

Radio galaxies are usually elliptical active galaxies that have two bright radio lobes. The energy that feeds the radio lobes comes from two jets emitted by the small nucleus located at the center of the elliptical galaxy, which shows the Cygnus A radio galaxy. The Cygnus A radio galaxy is the brightest radio object beyond the Milky Way. Radio and Seyfert galaxies seem to be powered by a supermassive central black hole. Seyfert galaxies are generally quiet at radio wavelengths.

Quasars

In the second half of the 20th century, astronomers discovered small and distant sources of radio radiation that were invisible in the optical part of the spectrum. Because the sources of radiation were small, astronomers called these objects **quasi-stellar radio sources**, or **quasars**. Quasi-stellar means star-like. Further work revealed that these radio sources also contain small but bright luminous optical objects. These objects were star-like in appearance, so they were called **quasi-stellar objects** (QSOs). Today, the name quasar is used to represent both the radio and optical quasi-stellar sources. The emission spectrum of the first optical quasar discovered, the 3C273, showed an unusual large redshift.

The large redshift indicates a huge recessional (radial) speed of 48,000 km/s. Recall from chapter three that the redshift **z**, is defined as

$$z = \frac{\text{change in wavelength}}{\text{rest wavelength}} = \frac{\Delta \lambda}{\lambda_0}. \qquad (10\text{-}2)$$

If we know the redshift **z** the recessional speed can be calculated using the following equation.

$$V_{radial} = \frac{\Delta \lambda}{\lambda_0} c = zc. \qquad (10\text{-}3)$$

Once we know the radial speed of recession, the distance is found using Hubble law.

Most quasars have redshift larger than one. If we use equation 10-2 to determine the speed of the quasar, we will find that it has a recession speed larger than the speed of light. We know that this cannot be possible because nothing moves faster than the speed of light.

For objects with large cosmological redshifts, the theory of relativity tells us that the radial speed and the redshift **z** are related by

$$\frac{v}{c} = \frac{(z+1)^2 + 1}{(z+1)^2 - 1} \qquad (10\text{-}4)$$

Example

A quasar has a redshift **z** = 6.4. What is radial speed of recession?

$$\frac{v}{c} = \frac{(6.4+1)^2 - 1}{(6.4+1)^2 + 1} = \frac{53.76}{55.76} = 0.96$$

This quasar is receding with a radial velocity equal to 6% the speed of light! Notice that the result can be written as **v** = 0.96 c.

The previous discussion tells us that a large redshift implies a large distance. One of the farthest quasars discovered has a redshift of 6.4 and a distance of about 12 billion ly. More than 450 quasars with redshifts larger than 4 have been discovered. Astronomers usually give the redshifts z of quasars instead of the distance. Quasars have a long look-back time and bring valuable information to us from when the universe was very young.

We should bear in mind that the galaxies are fixed in space, but the space is expanding and it carries the galaxies with it. As we mentioned earlier the rate of expansion of the universe is changing. However, the redshift z can be calculated because it only involves the ratio between the change in wavelength divided by the unchanged or reference wavelength. See equation 10-3. In any event, the fractional change of the redshift must be equal to the fractional change of the size of the universe **V** from the moment in which the photons were emitted. This is expressed in the following equation:

$$z = \frac{\Delta \lambda}{\lambda_0} = \frac{\lambda_{observed} - \lambda_{emitted}}{\lambda_{emitted}} = \frac{V_{observed} - V_{emitted}}{V_{emitted}}$$

Solving for $V_{observed}$, we obtain

$$V_{observed} = V_{emitted}(z + 1), \quad (10\text{-}5)$$

where $V_{observed}$ represents the actual size of the universe, and $V_{emitted}$ represents the size when the signal was emitted.

Example

A quasar is observed to have a redshift of 4.8. How many more times larger is the universe now than when the signal was emitted? Using equation 10-5, we derive:

$$V_{observed} = V_{emitted}(4.8 + 1) = 5.8\, V_{emitted}.$$

The universe is 5.8 times larger than when the signal was emitted.

Quasars seem to be located at the heart of very distant galaxies, and they appear to be powered by a supermassive black hole that has a large accretion disk. Even though the quasars are one of the most distant objects in the universe, they can be observed because they are extremely luminous, but the galaxies that contain them are faint and difficult to detect. Some quasars emit energy equivalent to 100 times the energy emitted by the Milky Way. Figure 10-16 shows a Hubble Space Telescope view of a quasar and its host galaxy. The host galaxy has a shell of stars around the quasar. See the insert to the lower right in Figure 10-16.

The galaxies that have stars organized in shells, or arcs, usually have been involved in mergers or collisions with other galaxies.

Q29. Quasar A has a recession speed of 75,000 km/s and quasar B has a recession speed of 25% the speed of light. Which quasar has the largest redshift?

a. A
b. B
c. both have the same redshift
d. not enough information is provided to decide

Q30. We know that quasars are far away because their emission spectra show large redshift.

a. True
b. False

Q31. The redshift of galaxies is a consequence of the expansion of space.

a. True
b. False

Q32. What is the speed of recession of a quasar that has a redshift z = 6.0?

a. 45% the speed of light
b. 100% the speed of sound
c. 96% the speed of light
d. 320,000 km/s

Q33. A quasar has a redshift of 6. How many more times larger is the universe now than when the signal from the quasar was emitted?

   a. 2 times

   b. 4 times

   c. the same

   d. 7 times

Q34. Some quasars emit energy equivalent to 100 times the energy emitted by the Milky Way.

   a. True

   b. False

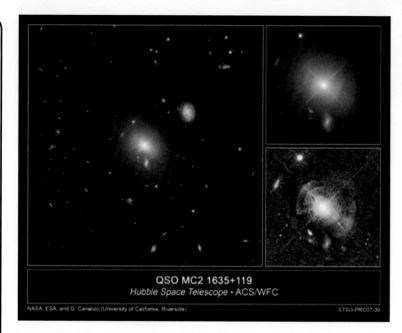

FIGURE 10-16. *The Hubble images show that the bright quasar, MC2 1635 + 119, resides inside a galaxy.*

Credit: NASA/ESA

QUASARS AND GRAVITATIONAL LENSING

In chapter nine, we saw that large concentrations of mass bend the light passing near it, creating what is known as a **gravitational lens**. If a galaxy is between us and a distant quasar, the light from the quasar is bent when it goes through the galaxy, forming an image of the quasar. See the diagram in Figure 10-17. The formation of multiple images of the same object due to a lensing galaxy is predicted by Einstein's general theory of relativity. Gravitational lensing allows astronomers to study distant objects. Quasars are objects with a very long look-back time.

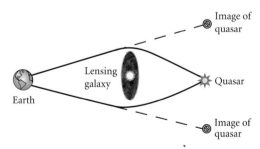

FIGURE 10-17. *A galaxy can produce images of distant objects. This is an example of a gravitational lens.*

A UNIFIED MODEL OF ACTIVE GALACTIC NUCLEI

Active galactic nuclei (AGN) includes Seyferts, radio galaxies, and quasars. All AGN are extremely luminous and most of the luminosity comes from small central and variable region. (The luminosity is due to the radiation emitted in different wavelengths, whether radio, infrared, visible, ultraviolet, and X-ray.) It seems that AGN are powered by an accretion disk around a supermassive black hole. According to this model, Seyfert and radio galaxies and quasars are the same thing. However, we see them as different objects because of the different orientation they have relative to Earth, and the different rate in which matter falls to their accretion disk.

The large amounts of energy that active galaxies emit come from the accretion disk around the central supermassive black holes that they have. Another important characteristics of the AGN is the variability of the energy that they emit in a short time. The luminosity of some quasars changes in week or days.

How were the black holes of the AGN formed? It seems that these supermassive black holes were formed as two galaxies merged. As the galaxies merged, large quantities of matter were dumped at the center of one of the intervening galaxies forming the supermassive black hole. Interactions between galaxies were more common when the universe was young because the galaxies were closer to each other than they are now. The activity of the AGN seemed to quiet down in about 100 million years or so, and then the galaxies become normal galaxies. Our own Milky Way and other galaxies might have been active galaxies in the past, but now their black holes are relatively quiet. AGN evolve and become normal galaxies.

Q35. AGN are powered by a supermassive black hole.
a. True
b. False

Q36. You discovered an object in the sky that you suspect might be a quasar. Which of the following observations will indicate whether it is a quasar?
a. A redshift $z = 4$
b. Extreme luminosity
c. Variable luminosity
d. It seems to be powered by a supermassive black hole
e. All the above

Galaxy Formation in the Early Universe

Even though light travels very fast (300,000 km/s), it has a finite speed. So the farther we look out into space, the longer the lookback time and the younger the part of the universe that we observe. Figure 10-18 shows the Ultra Deep Field image taken in March 2004, by the Hubble Space Telescope. The image shows distant and close galaxies. The larger spiral and elliptical galaxies are only 1 billion ly away, but the smaller galaxies are so distant that they were formed when the universe was only 800 million years

FIGURE 10-18. *Hubble's deepest view ever of the universe unveils earliest galaxies. Every image represents a galaxy.*

Credit: NASA Hubble space Telescope Collection

old. A portion of image 10-18 is shown in Figure 10-19. This image shows that distant galaxies do not have as definite structure as the closer galaxies.

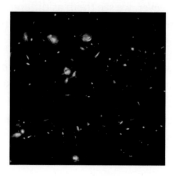

FIGURE 10-19. *Detail of the image of Figure 10-18.*

Credit: NASA Hubble space Telescope Collection

The observations made with the Hubble Space Telescope and other telescopes indicate that galaxies began to form from huge protogalactic clouds when the universe was young. These initial protogalactic clouds collided with each other in complete darkness. The collisions triggered the formation of the first stars and the first galaxies. The first stars formed when the universe was only 1 billion million years old. The first galaxies that formed were the spiral galaxies. These galaxies were smaller, bluer, and more irregular than the spiral galaxies that we see today. The elliptical galaxies appear to be the result of collisions between spiral galaxies. The collisions triggered star bursts that exhausted most of the gas and dust of the resulting galaxy. This is the reason why star formation is absent in elliptical galaxies. Galactic collision is a very slow process, and it lasts billion of years. As we said earlier, galactic collision was more common when the universe was young because the universe was smaller and the galaxies were closer to each other than they are now. However, collisions are still occurring, as shown in Figure 10-14.

Another interesting example of galactic collision is the Cartwheel Galaxy, located 500 million ly away in the constellation Sculptor. The Cartwheel Galaxy is the result of small objects to the right, passing perpendicularly through the disk of a larger spiral galaxy.

The Cartwheel Galaxy presumably was a normal spiral galaxy like our Milky Way before the collision. The collision rearranged the stars in a ring around the center of the galaxy. The Milky Way seems to be on a collision path with the Andromeda Galaxy. The collision might produce a large elliptical galaxies. No need to worry—the collision will happen in billions of years from now. The irregular and dwarf galaxies seem to be what remains of the collisions of larger galaxies.

ANSWERS

1. c
2. d
3. c
4. b
5. e
6. a
7. e
8. a
9. a
10. b
11. d
12. c
13. b
14. d
15. a
16. b
17. a
18. b
19. d
20. b
21. e, Note: The absolute magnitude is −19.6.
22. c
23. b
24. b
25. a
26. d
27. e
29. 25% the speed of light is 0.25 × 300,000 = 75,000, so both quasars have the same recessional speed; thus they have the same redshift.
30. a
31. a
32. c
33. d, See equation 10-5.
34. a
35. a
36. e

UNIT 5

In chapter ten we discovered that the universe begun around 13.7 billion years ago, with a Big Bang.

In this unit, we are going to study an experimental evidence that confirms the Big Bang theory, and also very briefly, we are to consider the different theories about the future of our universe.

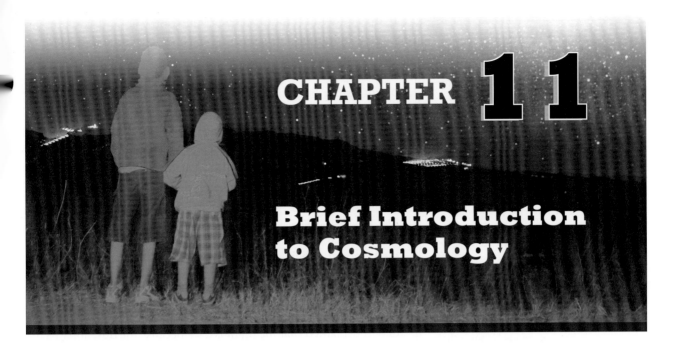

CHAPTER 11

Brief Introduction to Cosmology

We began our first encounter with astronomy by looking at the night sky and that is where we turn our eyes again toward the end of our journey. We ask ourselves, why is the sky dark at night? The answer is not trivial. In search of the answer, we will discover that the universe is not eternal and had a beginning.

In this chapter, we are concerned with a vision of the universe as a whole. How was it formed, and how it is evolving? The part of astronomy that answers these questions is the subject of cosmology.

THE COSMOLOGICAL PRINCIPLE

Two fundamental principles are the pillars of cosmology: homogeneity and isotropy. **Homogeneity** is the property of being uniform everywhere, and **isotropy** is the property of being approximately the same in all directions.

The cosmological principle tells us that universe is homogeneous and isotropic.

At first glance, the principle of cosmology contradicts our own experience in that the world is far from being homogeneous. Further, it does not look the same in all directions. For example, the plane of the Milky Way has more stars than the halo. The principle is valid only in very large scale such as superclusters of galaxies.

The cosmological principle implies that the universe does not have neither a center nor an edge. If there were a center, there would be at least one point different from any other point. If it had an edge, the points outside the edge will look different from the points that are inside the edge.

In summary, the universe is homogeneous and isotropic when it is observed on the scale of superclusters.

When Copernicus demoted the Earth from being the center of the universe, he was, in principle, stating that the universe does not have any privileged point of reference.

WHY IS THE NIGHT SKY DARK?

In 1823, the German astronomer Heinrich Olbert revived an old question, discussed previously by Kepler and others. Why is the night sky dark? This is known as **Olbert's paradox**.

> **Q1. The cosmological principle affirms that the universe is homogeneous. Using this information, which of the following is true?**
>
> a. Our solar system is far from being homogeneous so this principle is not true.
>
> b. Our universe is homogeneous only at the level of supercluster.
>
> c. The universe was homogeneous until it was 2 seconds old.
>
> d. The universe cannot be homogeneous because the space between the galaxies is empty.
>
> **Q2. When Copernicus removed the center of the universe from the Earth, he basically was saying that the _____.**
>
> a. the universe was isotropic
>
> b. the Earth was younger than the Sun
>
> c. the universe did not have any privileged position
>
> d. the universe was expanding

If the universe is homogeneous and isotropic and contains an infinite number of stars, then every imaginary line that we draw from Earth will end in a star, and every point in the night sky will receive light from an infinite number of stars causing it to shine as bright as the Sun. However, that is not the case. So what is the explanation? The problem is illustrated in Figure 11-1.

To look at the same problem in a slightly different way, let's place the Earth at the center of a sphere that contains the stars of the universe. See Figure 11-2. The number of stars that we see at a distance (**d**) will increase as the square of the distance (**d²**), and the brightness of each star will decrease by the inverse square of the distance (**1/d**)². We see that the total decrease in the brightness of the stars with distance is compensated by the number of stars at that distance. Therefore, the total amount of light received on Earth from the stars located at any given distance is the same. Thus, any point in the sky during the day or during the night will receive radiation from all the stars in the universe and will shine brightly. Then, why is the night sky dark?

Olbert explained that the night is dark because the dust absorbs the light from the stars. This is not a satisfactory answer. If the universe has existed forever, or for a very long time, the dust would have absorbed enough energy and become as bright as the surface of the stars.

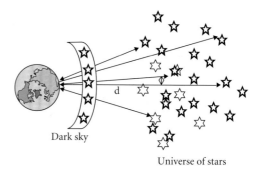

FIGURE 11-1. *If we get light from an infinite number of stars, the sky should shine brightly all the time.*

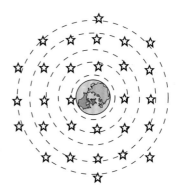

FIGURE 11-2. *The stars located on the surface of any of these hypothetical spheres contribute with the same amount of light to lighten our sky.*

So then, why is the night sky dark?

One possible explanation is that the expansion of the universe has redshifted the light from distant stars and galaxies to longer and longer wavelengths so that we cannot see them. This effect is part of the answer, but that is not enough. Calculations show that the reduction of energy due to the redshift is not enough to explain why the night is dark.

The night sky is dark because the universe has not existed forever. (See Big Bang in chapter ten.) We receive radiation only from the stars within a distance of 13.7 ly, because the universe is 13.7 billion years old.

The paradox is resolved when we accept that the universe has a beginning and it is continuously expanding.

Because the universe is not infinite and the light travels with a finite speed (300,000 km/s), there is a cosmic horizon. The space inside the cosmic horizon is the observable universe. The observable universe has a look-back time of 13.7 ly. The sky has not being receiving light from the stars for an infinite number of years. If that were the case, the sky would shine like the surface of our Sun. But that is not the case because the universe is not infinite in time.

THE FUTURE OF THE UNIVERSE

The theory of the Big Bang and the expansion of the universe, briefly mentioned in chapter ten, are currently accepted by most members of the scientific community.

The universe has been expanding since the Big Bang or for about 13.73 billion years. A question arises as to whether the expansion will come to an end and begin to collapse on itself (closed universe) or will it continue to expand for ever (open universe)?

Whether the universe is open or closed depends on the relation between the value of a parameter known as **critical density** and the actual or true density of the matter in the universe. (Remember that density results when we divide the mass of a given object by its volume V [density = mass/volume].)

The critical density is the average density that all the matter in the universe—normal and dark matter—should have in order to stop the cosmological expansion of the universe in the future.

The value of the critical density depends on the value of the Hubble constant. If H = 70 km/s/Mpcs, then the critical density is only 10^{-30} g/cm³. This roughly equates to about three atoms of hydrogen per each cubic centimeter. This low density corresponds almost to a perfect vacuum that cannot be achieved on Earth by any means. The value of the true density of the universe is difficult to obtain. However, it seems to be close to the value of the critical density.

Q3. The cosmological principle tells us that the universe _____.
 a. does not have an edge
 b. does not have a center
 c. is less than 10 billion years old
 d. a and b

Q4. Because the sky is dark at night, we can conclude that _____.
 a. the universe has been around forever
 b. dark matter exists
 c. the universe has a finite age
 d. dust blocks our view in almost every direction

Q5. The true density of the universe is close to 1 g/cm³.
 a. True   b. False

Q6. If the true density of the universe is larger than _____, it will eventually collapse on itself.
 a. the density of water
 b. the density of dark matter
 c. the critical density
 d. the density of the Milky Way
 e. 1 g/cm³

Q7. If the true density of the universe is less than the _____, it will expand for ever.
 a. the density of water
 b. the density of dark matter
 c. the critical density
 d. the density of the Milky Way
 e. 1 g/cm³

> **Q8. The "big crunch" theory about the universe implies that _____.**
>
> a. the expansion of the universe might reverse causing it to recollapse
>
> b. the universe will shrink in the far future, and when it is as large as the Milky Way, it will suddenly explode
>
> c. the universe will shrink to abut 50% of its present size and stop

Theoretical considerations show that if the density of matter is larger than the critical density, the universe would be closed. If the true density is smaller than the critical density, the universe would be open and infinite in extend. Finally, if the density is the same as the critical value, the universe is on the boundary between open and closed but still infinite in extension. Cosmologists state that the geometry of the universe is flat in this case.

A closed universe, or a high-density universe, has enough mass to slow down its expansion and eventually will come to a halt and collapse on itself. At some point in the future, according to this model, the entire content of the universe will accumulate in small volume forming a dense hot state. This state might trigger another Big Bang. This model is often called the **big crunch** and might repeat over and over in the future, as it might have happened countless of times in the past. In Figure 11-3, the evolution of a close universe is represented by curve 1.

The space in a closed universe curves around on itself, with positive curvature, making the universe finite in extension. This universe does not have a boundary even though it is finite in size. A two-dimensional representation of this universe is the surface of an expanding balloon where all objects are located.

In a closed universe, if you travel in a straight line always in the same direction, you will come back to the points of departure at some point in the distant future.

A low-density universe, or open universe, does not have enough matter to stop the expansion and will expand forever. The cluster of galaxies will drift away and eventually all of them will be invisible. The galaxies of the local group will remain approximately in the same place since the expansion of the universe is only observable at cosmological distances (at the level of superclusters).

In a two-dimensional analogy, the surface of an open universe is similar to the surface of a saddle and has negative curvature.

In this case, the evolution of the universe is represented by the two upper curves, as shown in Figure 11-3.

In the third case, when the two densities are equal, the geometry of the universe is flat because it does not have curvature. In this space, the Euclidean geometry that we are familiar with is applicable. For example, parallel lines will always remain parallel, while in a closed universe with positive curvature, parallel lines will intercept, and in an open universe, with negative curvature, parallel lines will diverge. Observations seem to confirm that the universe is flat.

The three different models briefly described are continuously expanding and all began with the Big Bang. These models just represent the different types of possible evolution of the universe.

A flat universe, or Euclidean universe, resembles a sheet in two dimensions. A closed universe has a spherical geometry and looks like the surface of a sphere. An open universe is hyperbolic and resembles a saddle-shaped surface in two dimensions. Each

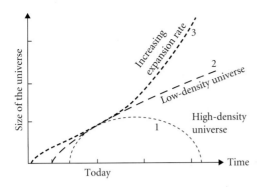

FIGURE 11-3. *The value of the density of the universe determines its rate of expansion and evolution.*

geometry offers an unbounded space that encompasses the entire universe.

It is important to emphasize that when we say that the universe is flat, we mean that it is Euclidean and that two parallel lines always remain parallel. (Draw two parallel lines on a sheet of paper, and make them as long as you want. They always remain parallel).

Recent observations of supernovae Type Ia indicate that the rate of expansion of the universe is accelerating. See curve 3 in Figure 11-3. This is a surprising result.

If the rate of expansion is accelerating, some force must be producing this acceleration. The cause cannot be normal matter or dark matter because the effect of normal and dark matter is to produce attraction not repulsion. The universe seems to contain "something" that produces a repulsive force that causes the acceleration of the expansion. This agent has been called **dark energy** or **quintessence**. This dark energy permeates the entire universe. The NASA Wilkinson Microwave Anisotropy Probe (WMAP) Explorer has determined that the universe contains 72.1% of dark energy and that its geometry is flat.

THE COSMIC MICROWAVE BACKGROUND

If the universe is expanding, there was a time in which it was very dense and hot. In fact, it had a temperature of trillions of degrees. In 1946, George Gamow, a Russian scientist who migrated to the USA, suggested that the universe had a temperature of more than a billion K during the early moments of its existence. Gamow's further work, along with that of his collaborators, showed that the expansion of the universe would cause its temperature to continually drop. For example, according to their model, the temperature of the universe was about 100,000 K 10 years after the Big Bang and 3,000 K after 380,000 years.

During the first 380,000 years after the Big Bang, the temperature of the universe was high enough to keep all the gases ionized. This

Q9. There is certain experimental evidence that the expansion of the universe _____.

a. has never changed

b. is decreasing

c. is increasing

d. is constant now, but in the future will slow down

Q10. The universe contains about _____ of dark energy.

a. 25%

b. 0%

c. 72%

Q11. The cosmic background radiation provides strong evidence that _____.

a. the early universe had a temperature of 2.75 K

b. the universe is isotropic

c. the universe has not changed since the Big Bang

d. hydrogen fuses into helium

Q12. The cosmic background radiation looks the same in any direction in the sky because _____.

a. the universe is closed

b. the expansion of the universe is increasing

c. the photons still retain the distribution they had before the last scattering

d. it is scattered uniformly by the Earth's atmosphere

Q13. It is believed that during the first 380,000 years after the Big Bang, the universe was opaque because the photons were efficiently scattered by free electrons.

  a. True
  b. False

Q14. What evidence do we have that the rate of expansion of the universe is increasing?

  a. the presence of dark matter
  b. the value of the critical density
  c. the CMB
  d. observations of supernovae Type Ia
  e. c and d

Q15. The universe is expanding into preexisting space.

  a. True
  b. False

Q16. Current accepted theories about the fate of the universe include _____.

  a. cosmic collapse
  b. infinite expansion
  c. a and b

Q17. When the temperature of the universe dropped to below 3,000 K _____.

  a. all atoms became ionized
  b. protons formed
  c. neutrons formed
  d. nuclei recombined with electrons to form atoms
  e. b and c

formed a plasma of nuclei and free electrons. In this hot plasma, the free electrons were very efficient at scattering the radiation (photons) produced by the hot and young universe. In this young and hot universe the photons did not travel far before they were scattered by the free electrons, which made the universe opaque. This early and hot universe was in equilibrium with matter since radiation did not escape. Matter and energy were constantly interchanging states, from matter to energy and from energy to matter.

This situation changed when the temperature of the universe dropped below 3,000 K. At this temperature, the nuclei combined with the electrons to form neutral atoms of mainly hydrogen and helium. As a result, the universe became transparent to radiation and the photons were able to escape and freely travel.

The photons that we see now were emitted at this time and have been traveling in the expanding universe for 13.7 billion years. The phase at which the universe became transparent is known as the **last scattering epoch** or **recombination era**. It is called the **last scattering epoch** because the electrons no longer scattered the photons and is called the **recombination era** because the nuclei captured electrons to become neutral atoms. Because matter and radiation practically stopped interacting after the last scattering epoch, this time period is also known as **matter-radiation decoupling**.

The photons (radiation) emitted during the last scattering carry the energy corresponding to an average temperature of 3,000 K and had a blackbody spectrum, peaking at a wavelength of 1,000 nm, as predicted by Wien's law.

$$\lambda = \frac{2,900,000}{3,000} \text{ nm} \approx 1,000 \text{ nm}$$

The radiation was emitted about 13.7 billion years ago and has been redshifted by the expansion of the universe. Since the last scattering, the universe has expanded about 1,000 times. According to equation 10-5, the radiation has also been redshifted 1,000 times, meaning that the radiation we receive from the last scattering has a maximum wavelength of about 10^6 nm and has blackbody distribution corresponding to an average temperature of 2.9 K. The value of the temperature is again obtained using Wien law.

If the universe is isotropic, as the cosmologic principle maintains, then the radiation emitted after the last scattering should be coming to us from all directions in the sky.

This radiation is known as the **cosmic microwave background** (CMB).

CMB we detect today was emitted after the last scattering at the moment of decoupling, and it has been redshifted by the expansion of the universe. (At the time of the last scattering the wavelength of the radiation was 1,000 nm and now it is one million nm).

The CMB was detected for the first time in 1965, by two Bell Telephone Laboratories researchers, Arno Penzias and

Robert Wilson. They discovered that the CMB was isotropic, had blackbody distribution, and had an average temperature close to the predicted value. Wherever these researchers aimed their instruments, they detected exactly the same radiation, proving that the radiation was isotropic.

In 1989, the Cosmic Background Explorer (COBE) measured in greater detail the microwave background radiation and confirmed its predicted properties.

COBE found that the radiation has a spread of wavelength between 1 cm and 0.5 mm and had a maximum average temperature of 2.725 K. This value is very close to the one found earlier by Penzias and Wilson (see Figure 11-4).

> **Q18. Observations made with the Hubble Space Telescope indicate that the early universe was isotropic.**
> a. True
> b. False
>
> **Q19. CMB is the _____.**
> a. radiation emitted by some massive black holes
> b. solution of Olbert's paradox
> c. radiation emitted by the Milky Way galaxy
> d. radiation formed when the universe was young
> e. radiation emitted by atomic hydrogen

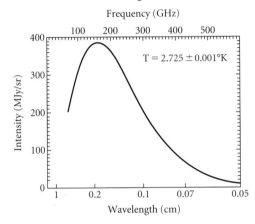

FIGURE 11-4. *Blackbody radiation of the cosmic background radiation.*

FLUCTUATIONS OF THE CMB AND THE FORMATION OF GALAXIES

COBE discovered that the CMB presents tiny variations in intensity, indicating that the isotropy of CMB is not perfect. These tiny variations in the intensity of the CMB indicate that the density of the universe, up to the last scattering, was not completely homogeneous but had small fluctuations. Several other experiments designed to measure these fluctuations have confirmed and refined COBEs result. Through the action of gravity, these tiny density fluctuations created the seeds that later on gave rise to the formation of galaxies, galaxy clusters, and the large-scale structure that we see in the universe today.

Astronomers have arrived at this conclusion because the small anisotropy in the CMB had to be present before and during the epoch of the last scattering. The CMB retains an almost perfect record of the universe when the CMB was emitted during the recombination era. If the early universe would have been

> **Q20. A low-density universe will continue to expand forever.**
>
>   a. True
>
>   b. False
>
> **Q21. COBE found the CMB to be absolutely uniform everywhere.**
>
>   a. True
>
>   b. False
>
> **Q22. The nuclei of heavy elements, like gold, were made during the early epochs of the universe before the last scattering.**
>
>   a. True
>
>   b. False
>
> **Q23. All the nuclei of hydrogen were made ____.**
>
>   a. soon after the Big Bang
>
>   b. after the first galaxies were formed
>
>   c. after decoupling
>
>   d. just before the Big Bang
>
> **Q24. Almost all the helium nuclei in the universe were formed during the first 15 min after the Big Bang.**
>
>   a. True
>
>   b. False

perfectly isotropic and homogeneous, galaxies would have not been able to form and you and me would not be here.

ORIGIN OF THE ELEMENTS

In previous chapters, we learned that the stars are factories of elements heavier than hydrogen. We also saw that the most abundant elements in the universe are hydrogen (H) and helium (He). Since the stars do not make hydrogen, where does it originate? Even though the stars make helium, calculations show that the present abundance of helium of 25% could not have been all made inside the stars. So, where was most of the primordial helium formed?

To answer these questions, look at the universe a few moments after it was created in the Big Bang. We do not know what happen at time zero of the Big Bang but we can come close to it.

As we have said earlier, during the first moment after the Big Bang, the temperature of the universe was so high that energy (radiation) was transformed into mass and mass back into radiation. This process is explained by the famous Einstein equation $E = mc^2$, as briefly explained in chapter eight.

The radiation filled the early universe, and it consisted mainly of gamma rays and very short wavelength X-ray photons. The transformation of energy into matter involved two high-energy gamma photons to create a pair of elementary particles and antiparticles and neutrinos. This process is known as **pair production**.

An antiparticle has the same characteristics of the normal particle except that it has an opposite charge. For example, there is the positron (e^+) with positive charge and the electron (e^-) with negative charge. The mass is the same for both particles. There are also protons–antiprotons, neutrons–antineutrons, quarks–antiquarks, and other more exotic particles.

A pair of particle–antiparticle annihilate each other when they come in contact, releasing high-energy photons. The first pair to form in the early universe were a quark and its corresponding antiquark. Soon after they were created, they annihilated each other, producing energy in the form of gamma photons. Quarks are the building blocks of neutrons and electrons. For reasons that are not yet known, a few more quarks were made than antiquarks. The quarks that survived combined to form the neutrons and protons. If the number of quarks were equal to the number of antiquarks, no other particles would have occurred and you would not be here reading this notes!

The electron–positron combination was made much later on when the temperature of the radiation was lower. This pair of particles also annihilated upon contact, and again there was an imbalance between the number of electrons and positrons, resulting in more electrons than positrons.

The photons created in the annihilation of the particles and antiparticles eventually became part of the microwave background.

Theoretical considerations suggest that for every 10 billion antiparticles, the number of normal particles exceeded the number of antiparticles by one. This was enough to form the universe that we view today.

The time of pair creation and annihilation did not last very long because the temperature of the universe quickly fell as it expanded. At the end of the first 3 minutes after the Big Bang, the temperature of the universe was about 1 billion K and protons and neutrons fused together to form **deuterium nuclei**, which is an isotope of hydrogen and consists of 1 proton and one neutron. **Helium three**, an isotope of helium with two protons and one neutron, was also formed in the same manner. Helium three is stable, and it constitutes only about 10^{-5} the total mass of the universe. Deuterium is very reactive, and it quickly converted into heavier nuclei, particularly helium. Some nuclei of lithium seven were also formed by a similar process.

By the end of the first 15 min, 25% of the mass of the universe consisted of helium nuclei. The nuclei of hydrogen were formed earlier when the protons were made. Recall that hydrogen consists of one proton and one electron. (A hydrogen nucleus is a proton.)

After the first few minutes of the Big Bang, the abundance of elements (nuclei) consisted of 75% of hydrogen and 25% of helium.

The formation of heavier elements than helium was produced by stellar nucleosynthesis over a billion years later when the first stars were formed.

The formation of light elements soon after the Big Bang is called **primordial nucleosynthesis** to differentiate it from the stellar nucleosynthesis of heavier elements in the stars.

The agreement of 1) the observed abundance of light elements, 2) the theoretical predictions of their formation, and 3) the detection of the CMB with the predicted properties gives good experimental support to the Big Bang theory.

As discussed earlier in this chapter, the recombination era in which the hydrogen and helium nuclei captured electrons to form neutral atoms happened much later, approximately 380,000 years after the Big Bang, when the temperature of the expanding universe fell below 3,000 K.

I would like to close these notes by challenging you to keep your eyes open, so you can understand the mysteries of our wonderful and beautiful universe. A word of warning to the wise: if you find your antiparticle or antiyou, stay away from it, so you will not be annihilated!

Q25. The tiny ripples in the background radiation found by COBE gave rise to the ____.

a. CMB
b. formation of quarks
c. formation of protons and neutron
d. formation of quarks and antiquarks
e. galaxy formation

Q26. Observations made with the WMAP indicate that ____.

a. the universe is about 24 billions years old
b. 71% of the universe is in the form of dark energy
c. the temperature in cosmic microwave background contains small anisotropies
d. b and c

Q27. Most of the deuterium formed in the first minutes after the Big Bang ____.

a. is still around today
b. formed hydrogen nuclei
c. was converted into energy
d. reacted to form helium
e. reacted to form iron and other heavier nuclei

Chapter 11—*Brief Introduction to Cosmology*

Q28. Neutral atoms were not able to form before the recombination area because ____.

a. the temperature of the universe at that time was too low

b. the protons were made after that epoch

c. there were not enough antiparticles

d. the temperature of the universe was too high

Q29. COBE observations revealed ____.

a. the dimensions of the universe

b. the average temperature of the CMB

c. small variations in the cosmic microwave background

d. the dark energy in the Milky Way

e. b and c

Q30. The present abundance of hydrogen and helium nuclei was formed when the first supernovae Type II occurred.

a. True

b. False

Answers

1. b		16. c	
2. c		17. d	
3. d		18. b	
4. c		19. d	
5. b		20. a	
6. c		21. b	
7. c		22. b	
8. a		23. a	
9. c		24. a	
10. c		25. e	
11. b		26. d	
12. c		27. d	
13. a		28. d	
14. d		29. e	
15. b		30. b	

APPENDIX A

Units of Length

1 meter (m) = 100 centimeters (cm)

1 millimeter = 0.001 m = 0.1 cm

1 kilometer (km) = 1,000 m

1 nanometer = 10^{-9} m

1 micron = 10^{-6} m

1 astronomical unit (AU) = 150 million km

1 light year (ly) = 9.5×10^{12} km = 63,200 AU

1 parsec (pc) = 3.26 ly = 206,000 AU

When we write the prefix "kilo" before a unit, it means one thousand; the prefix Mega means one million.

Example

10 kilo parsecs = (10 kpc) = 10^3 pc

2.5 Mega parsecs (2.5 Mpc) = 10^6 pc

APPENDIX B

Some Useful Constants

Speed of light = 300,000 km/s = 3×10^8 m/s

Hubble constant = 71 km/s/Mpc

The gravitational constant, G = 6.67×10^{-11} N $\times$ m²/kg²

Radius of the Earth = 6,378 km

Luminosity (L) of the Sun = 3.9×10^{26} Watts

Average surface temperature of the Sun = 5,380 K

Radius of the Sun = 696,000 km (7×10^5 km)

APPENDIX C

APPENDIX C

Orbital Data of the Planets in the Solar System

PLANET	SEMI-MAJOR AXIS (AU)	SEMI-MAJOR AXIS (×10⁶ km)	SIDEREAL ORBITAL PERIOD (TROPICAL YEARS)	SIDEREAL ORBITAL PERIOD (DAYS)	ECCENTRICITY	INCLINATION OF ORBITAL PLANE	AVERAGE ORBITAL VELOCITY (km/s)	INCLINATION TO THE ECLIPTIC
Mercury	0.387	57.9	0.241	87.96	0.206	7°0'19"	47.89	0
Venus	0.723	108.2	0.615	224.7	0.007	3°23'41"	35.03	177°18'
Earth	1	149.6	1	365.26	0.017	0°0'0"	29.79	23°27'
Mars	1.524	228	1.881	686.98	0.093	1°51'1"	24.13	25°12'
Jupiter	5.203	778.3	11.86	4333	0.048	1°18'17"	13.06	3°07'
Saturn	9.539	1427	29.46	10759	0.056	2°29'9"	9.64	26°44'
Uranus	19.18	2884	84.01	30685	0.047	0°46'23"	6.81	97°52'
Neptune	30.06	4521	164.8	60188	0.009	1°46'15"	5.43	29°34'
Pluto	39.53	5959	248.6	90700	0.248	17°7'30"	4.74	122°30'

Physical Data of the Planets

PLANET	RADIUS (km)	RADIUS (EARTH RADII)	MASS (EARTH MASSES)	MASS (kg)	DENSITY (g/cm³)	GRAVITY (EARTH = 1)	ALBEDO (%)	TEMPERATURE (K)	ESCAPE SPEED (km/s)
Mercury	2,439	0.38	0.0562	3.30×10^{23}	5.43	0.38	6	100–700	4.3
Venus	6,052	0.95	0.815	4.87×10^{24}	5.24	0.91	76	726	10.4
Earth	6,378	1	1	5.974×10^{24}	5.52	1	39	210–300	11.2
Mars	3,374	0.53	0.1074	6.419×10^{23}	3.94	0.39	16	190–310	5.0
Jupiter	71,492	11.19	317.82	1.899×10^{27}	1.33	2.54	51	110–150	59.5
Saturn	60,268	9.46	95.2	5.69×10^{26}	0.70	1.07	50	95	35.5
Uranus	25,559	4.01	14.5	8.66×10^{25}	1.30	0.9	66	58	2.1
Neptune	24,764	3.88	17.1	1.03×10^{26}	1.76	1.14	62	56	2.3
Pluto	1,151	0.18	0.0022	1.5×10^{22}	2.10	0.06	50	40	1.1

APPENDIX D

The 40 Brightest Stars

STAR	NAME	APPARENT MAGNITUDE	SPECTRAL TYPE	LUMINOSITY (SUN = 1)	DISTANCE pc (ly)
α CMa A	Sirius A	−1.46	A1 V	21	2.6 (8.6)
α Car	Canopus	−0.72	F0 II	1,300	96 (312)
α Boo	Arcturus	−0.04	K1.5 III	100	11.3 (36.7)
α Cen A	Rigil Kentaurus	−0.1	G2 V	1.4	1.3 (4.4)
α Lyr	Vega	0.03	A0 V	49	7.8 (25.3)
α Aur	Capella	0.08	G8 III	134	12.9 (42.2)
β Ori A	Rigel	0.12	B8 Iab	53,000	237 (773)
α CMi A	Procyon	0.34	F5 IV-V	6.4	3.5 (11.4)
α Ori	Betelgeuse	0.58	M2 Iab	13,000	131 (427)
α Eri	Achernar	0.50	B3 V pec	640	44.1 (144)
β Cen AB	Hadar	0.60	B1 III	9,300	161 (525)
α Aql	Altair	0.77	A7 V	10	5.1 (16.8)
α Tau A	Aldebaran	0.86	K5 III	150	20.0 (65.1)
α Vir	Spica	1.04	B1 V	1,600	80.4 (262)
α Sco A	Antares	1.09	M1 Ib	8,500	185 (603)
α PsA	Fomalhaut	1.15	A3 V	12	7.7 (25.1)
β Gem	Pollux	1.16	K0 III	31	10.3 (33.7)
α Cyg	Deneb	1.26	A2 Ia	53,000	990 (3226)
β Cru	Mimosa	1.28	B0.5 III	1.1	108 (352)

(Continues)

The 40 Brightest Stars (Continued)

STAR	NAME	APPARENT MAGNITUDE	SPECTRAL TYPE	LUMINOSITY (SUN = 1)	DISTANCE pc (ly)
α Leo A	Regulus	1.36	B7 V	150	23.8 (77.5)
α Cru A	Acrux	1.39	B0.5 IV	3,700	125 (408)
α Cen B		1.40	K4 V	0.4	1.3 (4.4)
ε CMa	Adara	1.51	B2 II	6,400	132 (431)
λ Sco	Shaula	1.6	B1 V	1,900	216 (703)
γ Ori	Bellatrix	1.64	B2 III	2,800	74.5 (243)
β Tau	Elnath	1.65	B7 III	300	40.2 (131)
β Car	Miaplacidus	1.70	A1 III	64	34.1 (111)
γ Cru		1.63	M4 III	230	27.0 (87.9)
ε Ori	Alnilam	1.70	B0 Ia	49,000	412 (1342)
α Gru	Al Na'ir	1.76	B7 IV	210	31.1 (101)
ζ Ori	Alnitak	1.79	O9.5 Ib	23,000	251 (817)
ε UMa	Alioth	1.79	A0p	59	24.8 (80.9)
α Per	Mirfak	1.82	F5 Ib	8,400	182 (592)
α UMa	Dubhe	1.81	K0 III	160	37.9 (124)
ε Sgr	Kaus Australis	1.81	B9.5 III	77	44.3 (145)
γ Vel	Suhail al Muhlif	1.83	WC8	37,000	258 (840)
δ CMa		1.85	F8 Ia	120,000	550 (1791)
β Aur	Menkalinan	1.86	A2 V	41	25.2 (82.1)
α Cru B	Acrux	1.90	B1 V	1,500	125 (408)
θ Sco	Sargas	1.86	F0 Ib	700	83.4 (272)

APPENDIX E

Some Local Group Galaxies

	TYPE	LUMINOSITY (SUN = 1)
M31 (Andromeda)	Sb	2.1×10^{10}
Milky Way	Sb/Sc	1.3×10^{10}
M33 = NGC 598	Sc	2.8×10^{9}
Large magellanic cloud	Ir	1.3×10^{9}
M32 = NGC 221	E2	2.8×10^{8}
NGC 6822	Ir	2.8×10^{8}
NGC 205	Spheroidal	2.6×10^{8}
Small magellanic cloud	Ir	2.3×10^{8}
NGC 185	Dwarf spheroidal	1.0×10^{8}
NGC 147	Dwarf spheroidal	8.5×10^{7}
IC 1613	Ir	7.0×10^{7}
WLM	Ir	3.4×10^{7}
Fornax	Dwarf spheroidal	2.3×10^{7}
And I	Dwarf spheroidal	4.1×10^{6}
And II	Dwarf spheroidal	4.1×10^{6}
Leo I	Dwarf spheroidal	3.7×10^{6}
DDO 210	Ir	3.1×10^{6}
Sculptor	Dwarf spheroidal	1.5×10^{6}
And III	Dwarf spheroidal	1.0×10^{6}
Pisces	Ir	9.3×10^{5}

(*Continues*)

Some Local Group Galaxies (Continued)

	TYPE	LUMINOSITY (SUN = 1)
Sextans	Dwarf spheroidal	7.7×10^5
Phoenix	Dwarf Ir/Dwarf spheroidal	7.0×10^5
Tucana	Dwarf spheroidal	4.9×10^5
Leo II	Dwarf spheroidal	4.4×10^5
Ursa Minor	Dwarf spheroidal	2.8×10^5
Draco	Dwarf spheroidal	2.1×10^5
Carina	Dwarf spheroidal	8.5×10^4
EGB 0427 + 63	Dwarf Ir	—
Sagittarius	Dwarf spheroidal	—

APPENDIX F

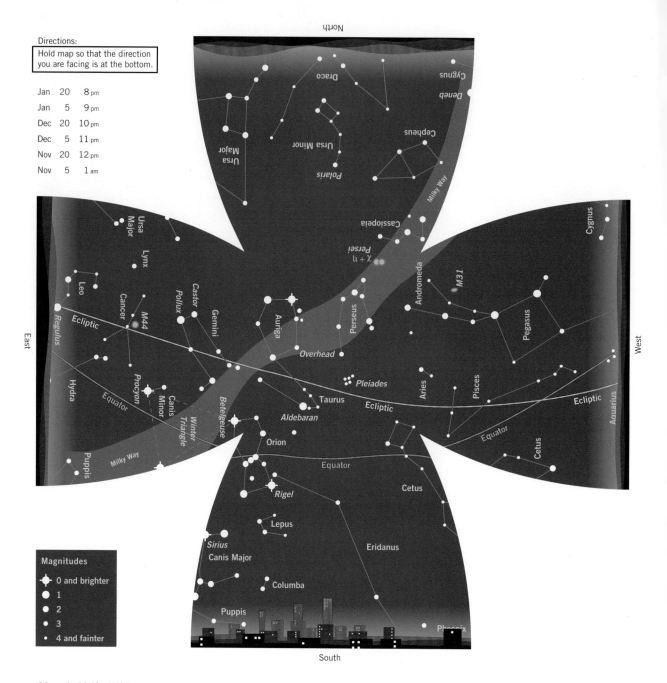

Map 1 [22^h–10^h]

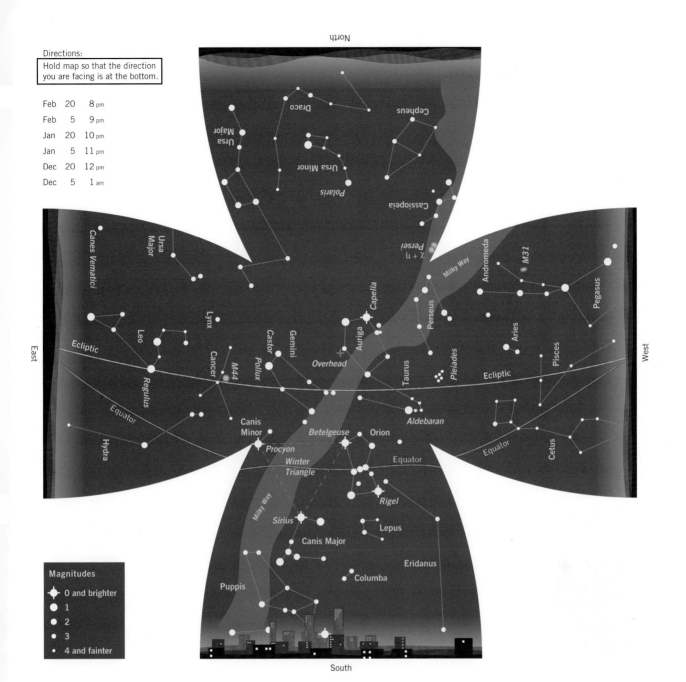

Map 2 [0ʰ–12ʰ]

From *Discovering Astronomy*, 5th Edition, by Stephen J. Shawl, Keith Ashman, and Beth Hufnagel. Copyright © 2006 by Kendall Hunt Publishing Company. Reprinted by permission.

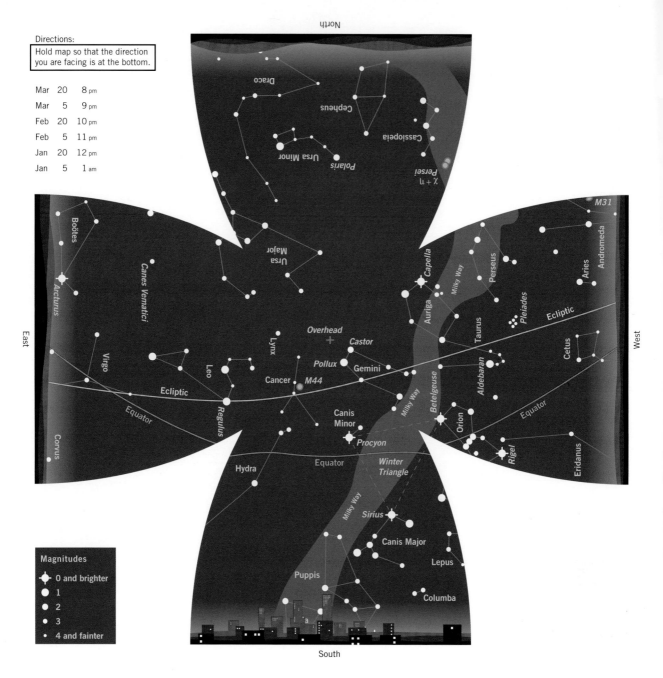

Map 3 [2ʰ–14ʰ]

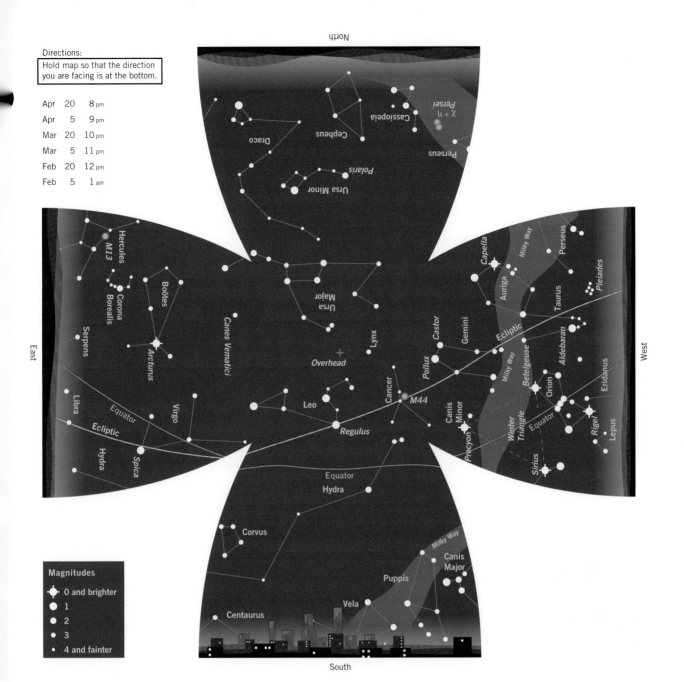

Map 4 [4ʰ–16ʰ]

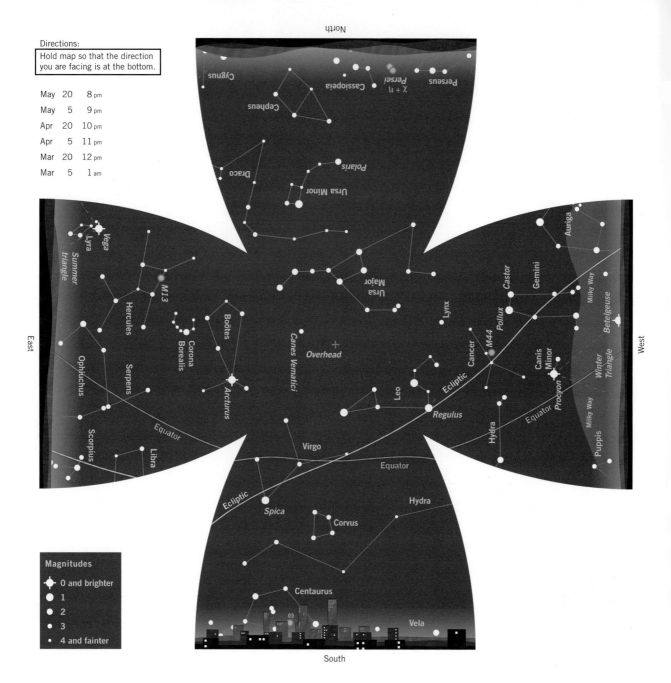

Map 5 [6ʰ–18ʰ]

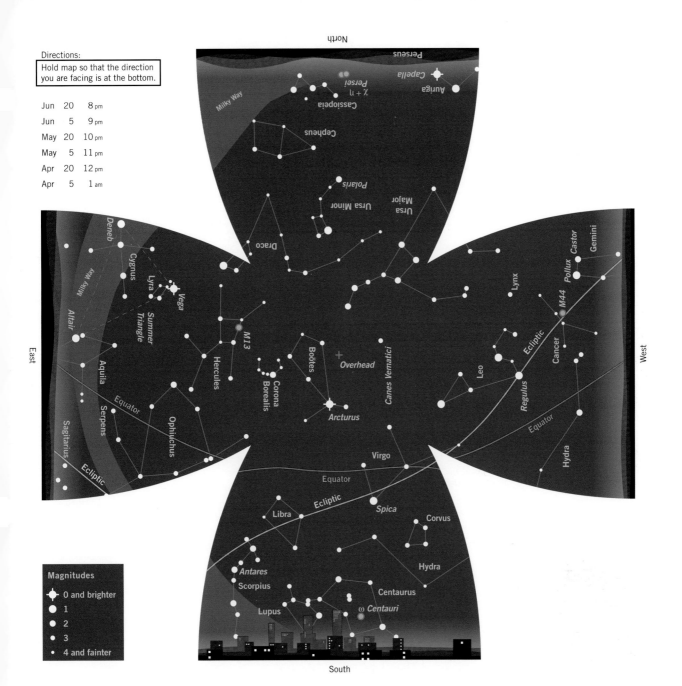

Map 6 [8ʰ–20ʰ]

Appendix F

Map 7 [10ʰ–22ʰ]

From *Discovering Astronomy*, 5th Edition, by Stephen J. Shawl, Keith Ashman, and Beth Hufnagel. Copyright © 2006 by Kendall Hunt Publishing Company. Reprinted by permission.

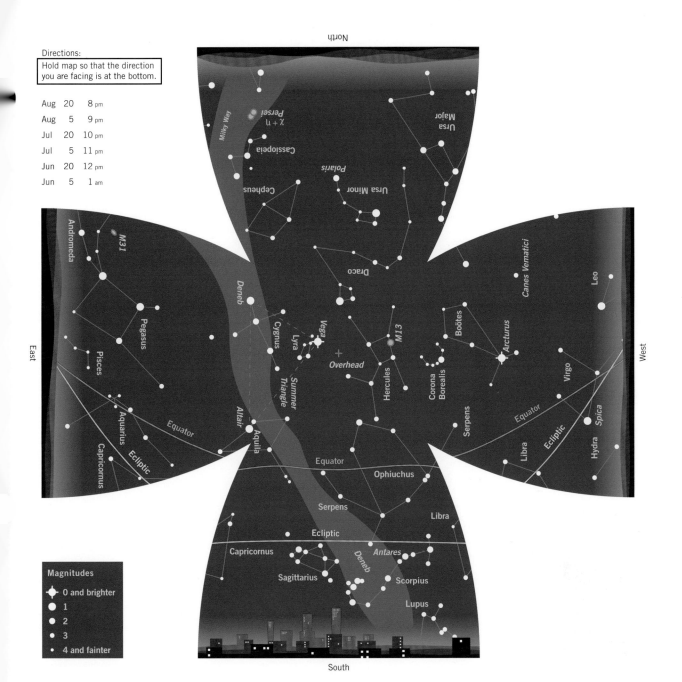

Map 8 [12ʰ–0ʰ]

From *Discovering Astronomy*, 5th Edition, by Stephen J. Shawl, Keith Ashman, and Beth Hufnagel. Copyright © 2006 by Kendall Hunt Publishing Company. Reprinted by permission.

Map 9 [14ʰ–2ʰ]

Appendix F

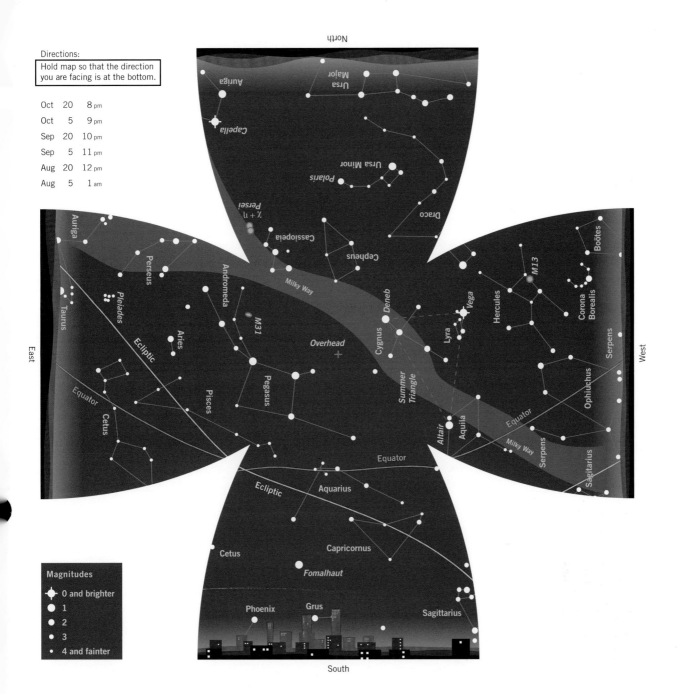

Map 10 [16ʰ–4ʰ]

From *Discovering Astronomy*, 5th Edition, by Stephen J. Shawl, Keith Ashman, and Beth Hufnagel. Copyright © 2006 by Kendall Hunt Publishing Company. Reprinted by permission.

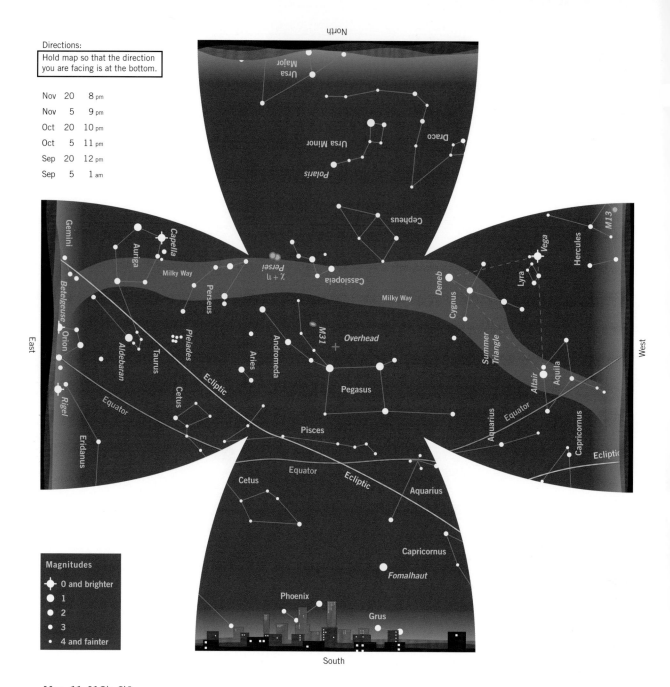

Map 11 [18ʰ–6ʰ]

Appendix F

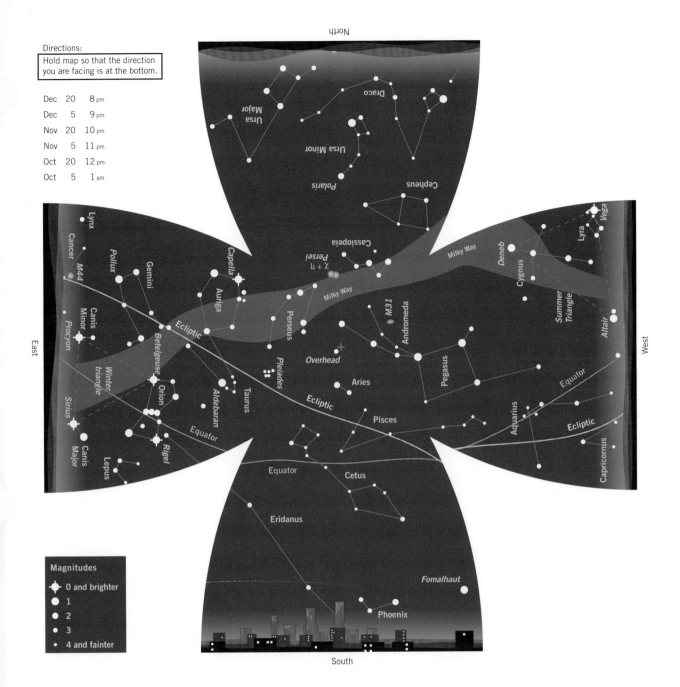

Map 12 [20ʰ–8ʰ]

APPENDIX G

Selected Web Sites Related to Astronomy

Hubble Space Telescope	http://hubblesite.org/
Chandra X-Ray Observatory	http://chandra.harvard.edu/
European Southern Observatory	http://www.eso.org/public/
National Optical Astronomy Observatory	http://www.noao.edu/
Spitzer Space Telescope	http://www.spitzer.caltech.edu/
NASA	http://www.nasa.gov/
Cassini Mission to Saturn	http://saturn.jpl.nasa.gov/
Galaxy Red Shift Survey	http://magnum.anu.edu.au/~TDFgg/
Astronomy Web Sites	http://www.stargazing.net/naa/sotw.htm
Astronomy Web Sites	http://faculty.physics.tamu.edu/allen/astronomy-websites.html
The Constellations	http://www.allthesky.com/constellations/const.html

APPENDIX C

INDEX

A

Absolute visual magnitude, 111
Absorption spectrum, 48, 52
Active galactic nuclei (AGN)
   definition, 213
   quasars
      definition, 214
      and gravitational lensing, 216
   radio galaxies, 213
   Seyfert galaxies, 213
   unified model of, 216–217
Adaptive optics (ADO)
   effect of atmospheric conditions, 68
   seeing conditions, 67–68
   uses of, 67
AGN. *See* Active galactic nuclei
Angular resolution
   power of telescopes, 65
   radio telescopes, 71
Angular separation, 20–21
Apollo asteroids, 87
Apparent brightness
   definition, 113
   vs. distance of stars, 113–115
Apparent visual magnitude, 111
Aristotelian geocentric model, 24
Asterism, 5
Atoms, 49. *See also* Hydrogen atom

B

Baryons, 49
Big Dipper, asterism, 5
Binary stars
   mass determination, 126
   types
      eclipsing binary stars, 127–129
      spectroscopic binary stars, 127
      visual binary stars, 126–127
Binding energy, 50
Blackbody radiators
   characteristics, 46
   Stefan–Boltzmann law, 48
   Wien's displacement law, 47
Black hole
   approaches, 186–188
   in binary systems, 188–189
   escape speed, 183–186
   event horizon of, 186–187
   formation, 187
   supermassive black holes, 187
Brown dwarfs, 142, 156

C

Cartwheel Galaxy, 218
CMB. *See* Cosmic microwave background
Cassegrian telescope, 61
Celestial axis, 5–6
Celestial equator, 6
Celestial sphere
   definition, 5
   elements
      celestial axis, 5–6
      celestial equator, 6
      circumpolar constellations, 7
      horizon and zenith, 6–7
Center for High Angular Resolution Astronomy (CHARA), 69
Cepheid variables stars, 172–173
Ceres, dwarf planet, 85, 87–89
Chandrasekhar mass limit, 163
Chandra X-ray observatory, 72
Circumpolar constellations, 7
Classical astronomy
   Aristotelian geocentric model, 24
   contributions of Greek astronomers, 23, 25
   Ptolemy's model, 26
Collapsar, 187

Comets
- Halley comet, 90
- long-period and short-period, 89–90
- nuclei of, 89

Compact objects, 164
Constellations, 3
Continuous spectrum, 52
Copernican heliocentric theory, 28
Core collapse supernova, 170
Cosmic Background Explorer (COBE), 197, 229
Cosmic microwave background (CMB)
- blackbody radiation, 229
- fluctuations, 229–230
- matter-radiation decoupling, 228
- photons, 227–228

Cosmological principle, 223

D

Dark matter, 195, 200–202, 212
Dark nebulae, 136
Deuterium, 231
Diffraction, 65
Diffraction limit, 65
Doppler effect, 55–58
Dwarf planets, 85
Dynamo effect
- definition, 85, 149
- in Jovian planets, 86

E

Earth-crossing asteroids, 87
Eclipses
- angular separation/small-angle formula, 20–21
- lunar eclipse, 16–17
- parallax, 19–20
- principle, 16
- solar eclipse, 18–19

Eclipsing binary stars, 127–128
Electromagnetic radiation
- definition, 41
- Earth's atmosphere, 43–44
- and electromagnetic spectrum, 41–43
- electromagnetic waves, 41–42
- temperature *vs.* heat energy, 44

Electromagnetic spectrum, 41–43
Electromagnetic waves, 41–42
Electrons, 49
Emission line spectrum, 48, 51, 53–54
Emission nebula, 55
- composition, 135
- definition, 135

Escape speed
- atmosphere of planets, 83–84
- compact objects and black holes, 183–186
- definition, 82

F

Failed stars, 142

G

Galactic disk
- arms, 196–197
- components, 195
- star formation, 196

Galaxies
- classification
  - elliptical galaxies, 204–205
  - irregular galaxies, 205–206
  - spiral galaxies, 204
- distance determination
  - redshifts of galaxies, 207–208
  - supernovae Type Ia, 206–207
- Hubble's law, 207
- interactions and collisions
  - look-back time, 212–213
  - mass and dark matter, 212
- supercluster, 210–211
  - 2dF Galaxy Redshift Survey (2dFGRS), 211
  - motion of galaxies, 211
  - voids, 211
- in universe, 208–210

Galex, 72
Galilean Moons, 32
Galilei, Galileo, 32
Gamma ray bursts (GRB), 188
Geocentric model of universe, 3
Giant Magellan telescope (GMT), 62
Giant molecular clouds (GMC), 94, 138
Giant planets. *See* Jovian planets
Giant stars, 120–121
Globular star clusters, 152
Gravitational lensing effect, 185
Gravitational redshift effect, 185
Greenhouse effect, 101–102

H

Halley comet, 90
Halo, 197–198
Heat energy, 43
Helium cycle, 160
Helium three, 231
Herbig-Haro objects, 144
Hertzsprung–Russell diagram
- giant stars, 120–121
- luminosity classes, 123–124
- main sequence (MS) stars, 120
- white dwarfs, 121

High mass main sequence stars
- core fusion and shell fusion, 168
- supernovae, 170–172
- surface temperature of, 167
- yellow Cepheids variables, 167

Hubble's law
- consequences, 208–209
- definition, 207

Hubble space telescope (HST), 73–74
Hydrogen atom
- binding energy, 50
- emission line spectrum, 51

Hypernova, 187

I

Interferometry. *See also* Optical interferometry
- definition, 68
- of radio telescopes, 70–71

Intermediate mass main sequence stars, 156–157

J

James Webb space telescope (JWST), 74
Jovian planets
  characteristics, 80–82
  formation
    atmospheres of, 100–101
    theories of, 99–100

K

Kepler, Johannes
  applications of, 30–31
  planetary laws of, 29–30
  modified third law, 35
Kitt Peak National Observatory (KPNO), 69

L

Large binocular telescope (LBT), 62
Light gathering power (LGP), 64
Lighthouse model, 179–180
Light spectra
  absorption spectrum, 52
  continuous spectrum, 52
  emission line spectrum/line spectrum, 53–54
  graphic representation, 54
  origin of hydrogen absorption lines, 52–53
Long-period comet, 89
Low mass main sequence stars, 155–156
Luminosity, 113
Lunar eclipse, 16–17
Lunar phases, 14–15

M

Main sequence (MS) stars
  characteristics, 142–143
  evolutionary track, 141
  intermediate/medium mass, 156–157
  lifetime of, 146
  low mass, 155–156
  luminosity, 128
  nuclear energy generation
    CNO cycle reaction, 147
    fusion of hydrogen liberate energy, 147
    proton-proton reaction, 146–147
  properties, 120
  protostars, 140–143
  stellar life-time, 145
Meteorites
  half-life, 93
  planetary impacts, 92
Metric system, 12–13
Milky Way
  age of, 199–200
  dark matter, 200–202
  galactic bulge, 197
  galactic disk, 196–197
  galactic nucleus, 198–199
  halo, 197–198
  mass of galaxy, 200–202
  motion of stars, 199
  origin of, 202–203
  structure of, 195
Millisecond pulsars, 182
Molecular clouds
  composition, 94
  dust grains, 94
  in equilibrium, 95
  and hydrostatic equilibrium, 138–139
  protostar, 139–140
Motion of earth
  orbital motion, 8–10
  precession, 11
  rotation, 8
Motion of moon
  lunar phases, 14–15
  rising and setting times, 15
  sidereal month, 14
  synchronous rotation, 14
  synodic month, 15
MS stars. See Main sequence stars

N

National Aeronautics and Space Administration (NASA), 72
Neap tides, 36
Near infrared telescopes, 71–72
Neutrons, 49, 169
Neutron stars (NS)
  in binary system, 181–182
  history, 176–177
Newtonian reflectors, 61
Newton's laws of motion, 34, 126
Newton's universal gravitation, 32–35
NGC 104 star cluster, 152
Nonstellar black holes, 187
Norma, 196–197
Nova, 164–167
Nucleosynthesis, 169

O

Olbert's paradox, 223
Open star clusters, 151–152
Optical/infrared interferometer, 69
Optical interferometry, 68–69
Optical telescopes
  giant Magellan telescope (GMT), 62
  instruments in, 62–63
  large binocular telescope (LBT), 62
  powers of
    angular resolution, 65–66
    light gathering power (LGP), 64
  reflectors, 61
  refractors
    drawbacks of, 60
    elements of, 59
  twin telescopes, 62
  very large telescope interferometer (VLTI), 62
Orbiting space-based telescope. See Hubble space telescope (HST)

P

Parallax, 19–20
Planetary laws of motion, 29–30
Planets
  dwarf planets, 85
  escape speed
    atmosphere of planets, 83–84
    definition, 82

Planets (*Continued*)
   Jovian planets
      characteristics, 80–82
      formation, 99–100
      magnetic fields, 85–86
   terrestrial planets
      characteristics, 78–80
      formation, 97–99
Population II stars, 195
Primordial nucleosynthesis, 231
Prominences, 150–151
Protons, 49
Ptolemy's model, 26
Pulsars
   definition, 178
   lighthouse model, 179–180
   synchrotron radiation, 179

Q

Quasars
   definition, 214
   emission spectra of, 214
   and gravitational lensing, 216
Quasi-stellar objects (QSOs), 214

R

Radial velocity, 57
Radiation laws. *See* Blackbody radiators
Radio galaxies, 213–214
Radio interferometer, 71
Radio telescopes (RTs)
   angular resolution of, 70
   elements of, 70
   interferometry of, 70–71
   resolving power, 70
Recurrent novae, 165
Red dwarfs, 155–156
Red giants, 158
Reflecting telescopes, 60
Reflection nebulae, 136
Refracting telescopes
   drawbacks of, 60
   elements of, 59
Resolving power. *See* Angular resolution
Rest wavelength, 56

RR-Lyrae variable stars, 160
Russell–Vogt theorem, 129

S

Sagittarius A* (Sgr A*), 198
Schwarzschild radius, 183
Seyfert galaxies, 213
Shock waves, 96
Shooting stars, 91, 92
Short-period comet, 90
Sidereal day, 10
Sidereal month, 14, 16
Sky at night
   asterism, 5
   constellations, 3–4
   dark, 223–225
Small-angle formula, 20–21. *See also* Angular separation
Solar and Heliospheric Observatory (SOHO), 149
Solar eclipse, 18–19
Solar flares, 150
Solar luminosity, 119
Solar nebula
   clearing, 102–103
   definition, 96
   hypothesis, 113
   temperature gradient, 97
Solar system
   characteristics, 78
   formation
      gravitational collapse, 96
      molecular clouds, 95
      temperature gradient, 97
   greenhouse effect, 101
   planetary nebula, 104
   planets
      dwarf planets, 85
      escape speed, 82–84
      Jovian planets, 80–82
      magnetic fields, 85–86
      terrestrial planets, 78–80
   properties of, 103–104
   solar nebula theory, 104
   space debris
      asteroids, 87–89
      comets, 89–90
      meteorites, 91–92
      meteoroids, 91
      meteors, 91–92
Solar wind, 86, 151

Space astronomy
   far infrared and UV telescopes, 72
   X-ray and gamma ray observations, 72–73
Space debris
   asteroids, 87–89
   comets
      Halley comet, 90
      long-period, 89
      nuclei of, 90–91
      short-period, 90
   meteorites
      half-life, 93
      planetary impacts, 92
   meteoroids, 91
   meteors, 91–92
Spectroscopic binary stars, 127
Spectroscopic parallax method, 112–113
Spring tides, 36–37
Star clusters
   globular clusters, 152–153
   open clusters, 151–152
   theoretical evolution of, 153
Stars
   apparent and absolute visual magnitude, 110–111
   apparent brightness
      definition, 113
      vs. distance, 113–115
   binary star system, 125
      eclipsing binary stars, 127–128
      mass determination, 126
      spectroscopic binary stars, 127
      visual binary stars, 126–127
   chemical composition, 118
   distance measurement
      spectroscopic parallax, 112–113
      stellar parallax, 109–110
   energy transport, 148
   Hertzsprung–Russell (H–R) diagram
      giant stars, 120–121
      luminosity classes, 123–124
      main sequence (MS) stars, 120
      white dwarfs, 121

interstellar medium (IM)
  composition, 133
  dark nebulae, 136–137
  emission nebula/HII regions, 135–136
  interstellar extinction, 134
  neutral atomic hydrogen detection/HI regions, 137–138
  reflection nebulae, 136
luminosity, 113
surface temperature, 116–117
T-Tauri stars
  binary system, 143–144
  bipolar flow, 144
  characteristics, 143–144
  Herbig-Haro objects, 144
Stefan–Boltzmann law, 48
Stellar black holes, 187
Stellar parallax method, 109–110
Stellar populations, properties of, 198
Sun
  composition, 151
  dynamo effect, 149
  Global Oscillation Network Group, 149
  prominences, 150
  solar flares, 150
  solar wind, 151
  structure, 149
  sunspots, 149–150
Sun-centered universe of Copernicus
  Copernican heliocentric theory, 28
  Galileo Galilei, 32
  Johannes Kepler
    applications of, 30–31
    planetary laws of, 29–30
  Kepler's modified third law, 35
  Newton's universal gravitation, 32–35
  Tycho Brahe, solar system, 28–29

Supercluster of galaxies
  2dF Galaxy Redshift Survey (2dFGRS), 211
  motion of galaxies, 211
  voids, 211
Supermassive black holes, 187, 199, 205, 217
Supernova remnant, 135, 171, 180
Supernova type Ia, 165–166, 207
Supernova type II, 170–171
Synodic month, 16

T

Temperature gradient, 97, 103
Temperature scales, 44–45
Terrestrial planets
  characteristics, 78–80
  formation
    accretion and planetesimals, 98
    condensation, 97–98
    density differentiation, 98–99
    protoplanets, 98
Thermal energy, 43
Tides
  consequences, 37
  definition, 35
  neap and spring tides, 36
Time dilation, 187
Trojan asteroids, 87
T-Tauri stars
  binary system, 143–144
  bipolar flow, 144
  characteristics, 143–144
  Herbig-Haro objects, 144
Twin telescopes, 62

U

Universe
  big crunch model, 226
  clusters of galaxies, 209–210
  critical density, 225
  dark energy/quintessence, 227
  expansion of, 208–209

galaxy formation, 217–218
local group of galaxies, 210
origin of elements
  composition of, 231
  electron–positron combination, 230–231
  pair production, 230
  particle–antiparticle annihilation, 230
  primordial nucleosynthesis, 231
Virgo cluster, 210
UV Ceti variables, 156

V

Vega, A0V star, 125
Very large array (VLA) interferometer, 71
Very large telescope interferometer (VLTI), 62, 69
Very long baseline array interferometer, 71
Virgo cluster, 210
Visual binary stars, 126–127

W

Weakly interacting massive particles (WIMP), 202
White dwarfs, 121
  interacting binary system, 164
  properties, 163–164
Wien's displacement law, 47

X

X-ray binaries, 164, 182, 189
X-ray bursters, 181
X-ray pulsars, 181. *See also* Pulsars

Z

Zero-age main sequence theory, 141